Adaptive and Personalized Visualization

Synthesis Lectures on Visualization

Editors
Niklas Elmqvist, *University of Maryland*
David S. Ebert, *Purdue University*

Synthesis Lectures on Visualization publishes 50- to 100-page publications on topics pertaining to scientific visualization, information visualization, and visual analytics. Potential topics include, but are not limited to: scientific, information, and medical visualization; visual analytics, applications of visualization and analysis; mathematical foundations of visualization and analytics; interaction, cognition, and perception related to visualization and analytics; data integration, analysis, and visualization; new applications of visualization and analysis; knowledge discovery management and representation; systems, and evaluation; distributed and collaborative visualization and analysis.

Image-Based Visualization: Interactive Multidimensional Data Exploration
Christophe Hurter
2015

Interaction for Visualization
Christian Tominski
2015

Data Representations, Transformations, and Statistics for Visual Reasoning
Ross Maciejewski
2011

A Guide to Visual Multi-Level Interface Design From Synthesis of Empirical Study
Evidence
Heidi Lam and Tamara Munzner
2010

Adaptive and Personalized Visualization

Alvitta Ottley

ISBN: 978-3-031-01479-6   paperback
ISBN: 978-3-031-02607-2   ebook
ISBN: 978-3-031-00351-6   hardcover

DOI 10.1007/978-3-031-02607-2

A Publication in the Springer series
SYNTHESIS LECTURES ON VISUALIZATION

Lecture #10
Series Editors: Niklas Elmqvist, *University of Maryland*
                David S. Ebert, *Purdue University*
Series ISSN
Print 2159-516X   Electronic 2159-5178

Adaptive and Personalized Visualization

Alvitta Ottley
Washington University in St. Louis

SYNTHESIS LECTURES ON VISUALIZATION #10

ABSTRACT

There is ample evidence in the visualization community that individual differences matter. These prior works highlight various *personality traits* and *cognitive abilities* that can modulate the use of the visualization systems and demonstrate a measurable influence on speed, accuracy, process, and attention. Perhaps the most important implication of this body of work is that we can use individual differences as a mechanism for estimating when a design is effective or to identify when people may struggle with visualization designs.

These effects can have a critical impact on consequential decision-making processes. One study that appears in this book investigated the impact of visualization on medical decision-making showed that visual aides tended to be most beneficial for people with high spatial ability, a metric that measures a person's ability to represent and manipulate two- or three-dimensional representations of objects mentally. The results showed that participants with low spatial ability had difficulty interpreting and analyzing the underlying medical data when they use visual aids. Overall, approximately 50% of the studied population were unsupported by the visualization tools when making a potentially life-critical decision. As data fluency continues to become an essential skill for our everyday lives, we must embrace the growing need to understand the factors that may render our tools ineffective and identify concrete steps for improvement.

This book presents my current understanding of how individual differences in personality interact with visualization use and draws from recent research in the Visualization, Human-Computer Interaction, and Psychology communities. We focus on the specific designs and tasks for which there is concrete evidence of performance divergence due to personality. Additionally, we highlight an exciting research agenda that is centered around creating tailored visualization systems that are aligned with people's abilities. The purpose of this book is to underscore the need to consider individual differences when designing and evaluating visualization systems and to call attention to this critical research direction.

KEYWORDS

visualization, individual differences, personality, cognitive abilities, user modeling

Contents

Preface

Society is saturated with data. This new access to information promises fresh opportunities for increased awareness, informed decision-making, and enhanced quality of life. Our greatest challenge now is making sense of it all, and visualizations have become *essential tools* for helping people explore, reason, and make judgments with data. One theory for why visual reasoning aides are useful in analytical environments is that visualization helps the analyst to externalize parts of the analytical process [105]. This fusion of human intuition with interactive visualization tools has proved to be invaluable for gaining insight into large or complex datasets [37].

The impact of visualization has begun to percolate into our daily lives. In recent years, there has been an explosion in public interest to collect data about themselves and their surroundings [149, 153–155], and to communicate their findings [167]. Media outlets such as *The New York Times* and *The Washington Post* are creating increasingly sophisticated visualization tools for readers to explore the data they collect. Although such tools promote transparency and provide an avenue for everyday users to experience data in new ways, prior work has shown that **data visualizations are not universally effective.**

There is overwhelming evidence in the visualization community that **individual traits matter**. In this book, we will dive into the literature that highlights how the distinguishing characteristics that identify one person from another, can influence how we reason with visual displays. Studies show that *personality* and *cognitive abilities* can modulate the use of visualization systems and demonstrate a measurable influence on speed, accuracy, process, and attention. In some circumstances, these effects critically impact important decision-making processes. For example, the work in Chapter 3 will investigate the impact of visualization on medical decision-making and show that visual aides tended to be most beneficial for people with high spatial ability, a metric that measures a person's ability to represent and manipulate two- or three-dimensional representations of objects mentally. The results will show that participants with low spatial ability had difficulty interpreting and analyzing the underlying medical data when they were presented with visual representations. We show that approximately 50% of the studied population were unsupported by the visualization tools when making a potentially life-critical decision. It is imperative to incorporate individual differences into design if we aim to make data visualizations more accessible to the diverse general public.

In addition, it is important to note that people with "advantageous" personality traits or cognitive abilities may also face problems when the data visualization tools they use are incompatible with their individual characteristics. For instance, one study shows that participants with high perceptual speed (those who compared objects quickly and accurately) suffer from lower accuracy in computing derived values if they use radar graphs instead of a heatmap [36]. Chapter 5 will show that people with internal locus of control, a measure of perceived control, were slower and less accurate with an indented tree visualization than with a dendrogram. Both high perceptual speed and internal locus of control correlated with high intellectual ability, and it is realistic to assume that many analysts possess these traits. Ultimately, these findings raise critical questions about whether one-size-fits-all visualizations are ideal and highlight flaws in design and evaluation practices.

This book presents my current understanding of *when* and *how* personality traits and cognitive abilities influence reasoning and decision-making with visualization tools. My goal is to present a view of how humans engage with data and draw from developments in fields such as human-computer interaction, visualization, and psychology.

WHERE WE ARE NOW

Experience, background, and abilities all shape the way we think and reason with data. Despite this fundamental truism, the majority of our visualization tools are still being built with a one-size-fits-all approach. This traditional approach to visualization design has mainly focused on mapping data to visual forms [14, 28, 106] and leveraging perceptual psychology to understand how people perceive these new designs [114, 160, 170, 171]. These theories, although essential, do not address how users think or how visualizations can be applied as an extension to an individual's cognitive ability. As a result, much like with many other user interfaces, visualization users are often forced to adapt their reasoning and analysis strategies to the tools' design.

This approach to visualization design is fast becoming inadequate [176]. Visualizations have emerged as an integral part of data analysis and decision-making, and often serve as an extension to the analyst's cognition. They are now being used to solve high-impact problems in many areas including health, business, and military, and an analyst will typically interact with one or more visualization to explore or reason about large and complex data. In order to support this process, it is ever important to *know the user*.

Realizing this, a new area of research has emerged in the Visualization community. Researchers have demonstrated how individual differences such as perceptual speed [3, 34, 36, 157, 158], spatial ability [27, 30, 90, 127, 165, 166, 182], experience [34, 47, 101], and other personality traits [21, 66, 67, 119–121, 128, 181, 185] can significantly affect how users approach a problem and their ability to use visualization tools. Most importantly, they highlight that there is no single visual design that suits every user.

This research is still in its infancy and has primarily focused on correlating individual differences to speed and accuracy with visualization tools. While these do highlight differences between groups of users, they give us little insight into *how* or *why* groups of users differ. Many open questions remain.

1. What are the most important traits to measure?
2. Do user groups adapt different problem-solving strategies?

3. How can we choose between existing tools?
4. How can we design new tools to better facilitate users' needs?

THE STRUCTURE OF THIS BOOK

This book aims to address some of these questions by providing a summary of key findings on individual differences in visualization use. The goal is to contextualize the prior works that span multiple fields and provide useful and practical tips. Admittedly, we will only scratch the surface of an intriguing and potential transformative research direction. However, this book hopes to be the catalyst for new and exciting innovations in personalized and adaptive visualizations.

Part I summarizes the fundamental definitions and concepts. In Chapter 2, we will begin with a survey of known cognitive factors that influence visualization use. Chapter 1 provides one possible framework for structuring the space of individual cognitive differences.

In Part II, we highlight studies that demonstrate the correlation between individual characteristics and visualization use. Chapter 3 provides empirical evidence for the importance of considering individual differences in a high impact real-world application domain. It demonstrates how *spatial ability* influences performance with visualizations for medical decision making. Chapter 4 explores other individual differences and identifies the trait *locus of control* as a key factor for predicting performance on various hierarchical visualizations. In Chapter 5, we take a step beyond speed and accuracy measures, and we examine how individual differences impact user search strategies. It explains *how* and *why* groups of users differ through a manual analysis of interaction log data.

Part III explores the feasibility of adaptation. Chapter 6 builds on the work in Chapter 5 by demonstrating how we can use machine learning techniques to automatically infer user attributes. It also investigates the viability of real-time adaptation. Chapter 7 suggest that adaptation can be a two-way construct in which both the user and computer adapt to each other. We demonstrate how psychological priming techniques can temporarily influence cognitive state and performance.

A concluding chapter, Chapter 8, explores the implications individual differences and discusses open challenges for designing next generation visualization tools.

Alvitta Ottley
March 2020

Acknowledgments

This book represents almost a decade of research, hundreds of manuscripts that spans multiple disciplines, and collaborations that created life-long friendships. Thank you to Remco Chang and Robert Jacob, who mentored and supported me. The work in this book would not be possible without Caroline Ziemkiewicz, Evan M. Peck, R. Jordan Crouser, Eli T. Brown, Daniel Afergan, Lane Harrison, Beste Yuksel, Paul Han, Holly A. Taylor, and Zhengliang Liu.

I am fortunate to have a family who supports and encourages me. Thank you Chizoba and Titan for your patience, love, and motivation.

Alvitta Ottley
March 2020

Text Credits

Chapter 1 is based, in part, on Evan M. Peck, Alvitta Ottley, Beste F. Yuksel, Remco Chang, and Lane Harrison. ICD 3: Towards a 3-dimensional model of individual cognitive differences. *ACM BELIV'12: Beyond Time and Errors: Novel Evaluation Methods for Information Visualization*, 2012. DOI: 10.1145/2442576.2442582 [130].

Chapter 3 is based, in part, on Alvitta Ottley, Evan M. Peck, Lane T. Harrison, Daniel Afergan, Caroline Ziemkiewicz, Holly A. Taylor, Paul K. J. Han, and Remco Chang. Improving Bayesian reasoning: The effects of phrasing, visualization, and spatial ability. *IEEE Transactions on Visualization and Computer Graphics*, 22(1):529–538, 2015. DOI: 10.1109/tvcg.2015.2467758 [126].

Data are available at: http://github.com/TuftsVALT/Bayes

Chapter 4 is based, in part, on Caroline Ziemkiewicz, Alvitta Ottley, R. Jordan Crouser, Ashley Rye Yauilla, Sara L. Su, William Ribarsky, and Remco Chang. How visualization layout relates to locus of control and other personality factors. *IEEE Transactions on Visualization and Computer Graphics*, 19(7):1109–1121, 2012. DOI: 10.1109/tvcg.2012.180 [184].

Chapter 5 is based, in part, on Alvitta Ottley, Huahai Yang, and Remco Chang. Personality as a predictor of user strategy: How locus of control affects search strategies on tree visualizations. In *Proc. of the SIGCHI Conference on Human Factors in Computing Systems*. ACM, 2015. DOI: 10.1145/2702123.2702590 [128].

Chapter 6 is based, in part, on Eli T. Brown, Alvitta Ottley, Helen Zhao, Quan Lin, Richard Souvenir, Alex Endert, and Remco Chang. Finding Waldo: Learning about users from their interactions. *IEEE Transactions on Visualization and Computer Graphics*, 20(1):2, 2014. DOI: 10.1109/tvcg.2014.2346575 [21].

Chapter 7 is based, in part, on Alvitta Ottley, R. Jordan Crouser, Caroline Ziemkiewicz, and Remco Chang. Priming locus of control to affect performance. In *IEEE Conference on Visual Analytics Science and Technology (VAST)*, pp. 237–238, 2012. DOI: 10.1109/vast.2012.6400535 [120].

PART I

Fundamentals

CHAPTER 1

A Framework for Thinking About Individual Differences

"[W]e should come up with ways to narrow down the number of survey questions and/or ways to select proper measures out of the comprehensive set depending on tasks and visualization techniques"

Ji Soo Yi   [176]

In a position statement presented at the *evaluation and BEyond—methodoLogIcal approaches for Visualization* (BELIV) workshop in 2010, Nathalie Riche [139] proposed the use of multiple physiological measurements (heart rate, eye gaze, brain imaging, etc.) for evaluating visualizations. At the same BELIV workshop, Ji Soo Yi [176] proposed studying individual differences when evaluating visualizations. Yi argued that understanding how users differ in personality and cognitive factors is important in evaluating visualizations. In a follow-up research project, he demonstrated that there is a significant difference between novice and expert users when using a visualization to solve analytical tasks and pinpoints the importance of additional research in individual differences in visualization evaluation [96].

The emergence of this body of research ultimately highlights the need for better evaluation methods that address the unique needs of visualizations, but there is no consensus on which methods address these needs. What is clear, however, is that the field of visualization is yet have a systematic and objective way of measuring individual differences in user analysis of visualizations. The absence of such a framework makes it difficult to better understand how factors in individual differences interact with each other and with existing evaluation techniques.

1.1   COGNITIVE STATES, TRAITS, AND EXPERIENCE

In this chapter, we present a first-step toward a solution by introducing the ICD^3 Model (Individual Cognitive Differences)—a three-dimensional model that encompasses the cognitive facets of individual indifferences in visualization use. A necessary step in attempting to define a model of individual cognitive differences was to seek an underlying structure of previous research by identifying which factors are dependent and which are independent of one another. By surveying the existing literature, my colleagues at Tufts University and I proposed that individual differences can be categorized into three orthogonal dimensions: cognitive traits, cognitive states and experience/bias [130].

Cognitive traits are user characteristics that remain constant during interaction with a visual analytic system. Factors such as color deficiencies, personality, spatial visualization ability, and perceptual speed are all examples of cognitive traits. These have been shown to correlate with a user's ability to interact with a visualization [27, 36, 66, 165, 185] and can be generalized to predict the behavioral patterns of users with different cognitive profiles.

Cognitive states, on the other hand, are the aspects of the user that may change during interaction and include situational and emotional states, among others. Research has shown that a user's performance can be significantly altered by changes in their emotional state [9, 54, 92, 132, 142, 147], and the importance of combining workload with performance metrics has been noted for decades [75, 129, 175]. Although cognitive states are difficult to measure because of their volatility, they provide important contextual information about the factors affecting user performance that cannot be described through cognitive traits alone.

Cognitive states and traits can describe a significant portion of a user's cognitive process but they are not comprehensive; experience and biases can also affect cognition. Intuitively, we think of experience and bias separately. However, they both describe learned experiences that can affect behavior when familiar problems arise, and are therefore not orthogonal. There has been growing interest in the visualization community to systematically measure and evaluate *Visualization Literacy*—a measure of visualization experience for non-experts [5, 17, 19, 102]. Previous studies have shown that cognitive biases can significantly affect performance and decision-making with visualization interfaces [44, 45, 49, 168, 169].

Altogether, these three dimensions can create a model that encapsulates the cognitive aspects of individual differences (Fig. 1.1). Similar to how analyzing state and trait alone would disregard potential performance gains from expertise, ignoring any one dimension of the model would also result in an incomplete description of performance. For example, analyzing only expertise and traits ignores changes that may be triggered by workload or emotions (cognitive state). Thus, the model is only complete if all three dimensions are considered. By using ICD^3, evaluators can identify what factors must be controlled in an experiment and which should be included as independent variables. The community can also begin to evaluate visualizations using this common platform and be able to better reproduce and extend each other's research.

1.2   THE ICD^3 MODEL

In light of the three dimensions that we have discussed, we believe that a structured model would be useful in describing individual cognitive differences when users interact with visualizations. In this model, each orthogonal dimension would represent an individual difference of a user thereby allowing researchers to describe or perhaps even predict a user's ability to interact with a visualization, by knowing where that individual lies along the three different axes. This would

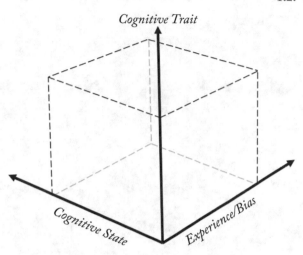

Figure 1.1: The ICD³ categorizes individual cognitive differences in three orthogonal dimensions: Cognitive Traits, Cognitive States, and Experience/Bias.

allow for not just isolated cognitive factors, but for the *interaction* of the user's different cognitive abilities.

Figure 1.2 gives a hypothetical example of a user looking at percentage judgments in treemaps. The cognitive state in this example is the user's workload, their cognitive trait is their working memory capacity, and their experience/bias is how experienced they are with treemaps. An ICD³ model would show that if the user is an expert, has a low workload, and has a high working memory capacity, then they have higher performance and abilities with percentage judgment in treemaps. Conversely, if the user is overloaded with work, has a naturally low working memory capacity, and has no experience of treemaps, then they will be less effective in performing that task.

After defining the visualization, task, and cognitive factors, a set of experiments can then be run in which participant workload, working memory capacity, and experience is varied. For each interaction of factors, performance is recorded in the instance at the appropriate coordinates. Given enough data, we construct a descriptive topology of performance for a task and visualization.

Unfortunately, the interaction of cognitive facets is ordinarily much more nuanced than depicted in Fig. 1.2. For the sake of simplicity, we chose working memory capacity, workload, and experience because their impact on performance is relatively straightforward. But in practice, we have little knowledge of how other combinations of state, trait, and experience/bias influence interaction with a visualization.

For example, some studies have shown that extraverts and introverts perform differently when they receive positive or negative feedback about a task, thus modifying their cognitive

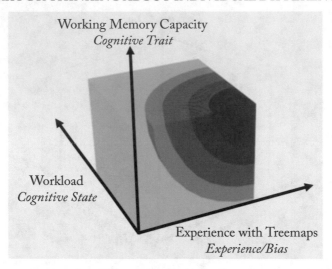

Figure 1.2: An example of how an ICD^3 model might be constructed. We map the interaction of workload, working memory capacity, and experience on performance of percentage judgments in treemaps. Darker red represents better performance.

state [16]. Introverts tend to perform well when given positive feedback and worse when given negative feedback. Reciprocally, extraverts perform worse than intraverts given positive feedback, but their performance improves under negative feedback. This exemplifies why it is important to consider the interaction of state and trait.

However, other studies have suggested that people with an external locus of control (LOC), which is correlated with extraversion [117], perform better in visualizations where they have had no previous experience than people with an internal LOC [185]. This demonstrates how trait and experience can interact to influence performance.

Each of these examples provide a two-dimensional snapshot of how cognitive dimensions can impact performance. But how do we combine the knowledge of these two studies? How would performance be impacted when an experienced intravert is given negative feedback, or an inexperienced extravert is given positive reinforcement during a task? Thus, a key attribute of the ICD^3 model is that limiting the scope of evaluation to any two of the three described dimensions leaves an incomplete and potentially misleading model of performance.

- Analyzing state and trait without experience ignores performance gains by expertise.

- Analyzing state and experience without trait ignores interaction differences that are driven by personality or inherent cognitive strengths (e.g., spatial ability).

- Evaluating experience and trait without state disregards the moment-to-moment cognitive changes in the user that could be driven by emotion or workload.

While instances of the ICD^3 model should be constructed for a explicit task and visualization, we imagine that the interaction of certain cognitive factors will be generalizable across visual forms (and tasks). In the next section, we explore the implications of the ICD^3 for design.

1.3   LIMITATIONS AND FUTURE WORK

Creating a precise model of individual differences is a daunting task. From the literature, we see that even the slightest deviations between people can influence performance on a visualization. Cognitive states may interact with and manipulate each other—for example, emotional state has been shown to impact working memory—and people simultaneously bring many traits and experiences to the table each time they see a visualization. Furthermore, there are almost certainly cognitive traits, states, and experiences that impact interaction significantly more than others.

While we do not believe that these problems impact the orthogonality of our proposed model, it illuminates the potential dependency of factors within each dimension, increasing the difficulty of predicting human interaction. We highlight at least two future areas of research that will be critical to addressing these challenges.

First, finding dominant individual cognitive factors both within dimensions and between dimensions should limit the sheer volume of cognitive tests necessary to describe interaction. For example, if participants have a low working memory capacity, their LOC might not matter given a certain task and a visualization. If this is true, then detecting and adapting to LOC may be unnecessary. Similarly, we suspect that a person's experiences and biases may impact performance more than many other cognitive traits and states. Thus, if we know a person is an expert at a simple task, emotional state might be irrelevant. Identifying these dominant factors should reduce the number of interactions between cognitive factors.

Second, discovering new and unobtrusive methods to capture cognitive state, trait, and experience/bias will ultimately drive research in individual cognitive differences. For example, in future chapters (Chapter 6), we will explore how we might detect user attributes by analyzing their click steam data. Recent advances in non-intrusive physiological sensors that detect emotional states, such as the Affectiva Q-Sensor [136], will enable future studies into the impact of emotional state and visualization performance. In real-world scenarios, it is unrealistic to expect users to be subjected to a deluge of forms and intrusive monitoring equipment. The simple act of filling out personality surveys or applying brain sensing equipment is enough to potentially modify cognitive state (or introduce biases) before interaction. It should be a central goal to develop new and effective ways in which we can automatically detect these states.

The generalizability of cognitive states, cognitive traits, experiences/biases on performance in visualization has yet to be seen. As a result, the ICD^3 model takes a conservative approach by specifying an exact set of cognitive factors and requiring tests to be performed on a fixed task and fixed visualization. By identifying important factors or important interactions between factors, we hope to construct new metrics in the future that are more predictive of interaction with a visualization.

1.4 SUMMARY

This chapter makes initial steps toward a model that captures the various cognitive aspects that affect visualization performance by dividing them into three dimensions: cognitive states, cognitive traits, and experience/bias. Furthermore, we have discussed how each of these dimensions are orthogonal to each other, meaning that during visualization interaction, a user may exhibit different values for states, traits, or experience/biases. Each of the dimensions are found to influence cognitive processes related to visualization, such as reasoning and perception. The ICD^3 model provides a sample space for experiments involving visualizations, so that we may form a better understanding of the cognitive underpinnings of visualization.

CHAPTER 2

Cognitive Traits that Matter

"According to both common sense and the assumptions of most employers, people vary, thus any group of workers performing virtually any type of task are not likely to be performing equally effectively at all times."

Andrew Dillon and Charles Watson   [43]

From Chapter 1, we see that individual differences are multidimensional. Experience and bias describe our evolving and learned behavior. Cognitive states express our situational condition at a specific time. Cognitive traits reveal stable factors that distinguish one person from another, and is arguably to most important of the three dimensions. As such, numerous studies have focused on identifying cognitive traits or "stable tendencies to respond to certain classes of stimuli or situations in predictable ways" [43]. These studies find their roots in Psychology, and research have shown that people with distinct personality types and cognitive abilities exhibit observable differences in task-solving and behavioral patterns [2, 174]. Studies dating back to the early 1900s began by investigating variation in workplace performance [76]. Throughout the intervening century, researchers have extended these findings to investigate individual characteristics that may predict performance under a variety of conditions.

Researchers in the computational sciences have also recognized the role that cognitive traits might play in shaping interaction in human-machine systems. For example, Benyon and Murray observed a relationship between spatial ability and task performance as well as user preferences when using common interfaces such as command line and menus [13]. In more recent studies, the effects of personality traits in addition to cognitive abilities have been of particular interest. Gajos and Chauncey [58] observed that introverted people were more likely to use adaptive features in user interfaces as compared to extroverts. Orji et al. [118] showed that conscientious participants responded well to persuasive strategies such as self-monitoring and feedback in gamified systems. These studies are just a small sample of a large body of work in the HCI community that documents the influences of personalities and cognitive abilities on their interactions with computer interfaces. For more complete surveys of the literature, see [10, 43, 135].

There is a growing interest in extending these findings to the field of data visualization. In particular, researchers have gradually but steadily realized the deficiency of having only one-size-fits-all data visualization interfaces as well as the importance of cognitive traits in the use of data visualization systems [176]. The most common approach in the visualization community is to classify people based on broad categories of user attributes such as personality traits and cognitive abilities, and identify the corresponding stimulus or situational variable. Research shows that personality and cognitive abilities could have substantial impacts on task performance [66, 181], usage patterns [21, 128] and user satisfaction [93] with various data visualization designs. Building on these findings, researchers have begun to examine how we might leverage cognitive traits for applications such as user modeling [21, 128] and adaptive interfaces [98].

This chapter provides an overview of the most frequently studied cognitive traits that researchers have linked to visualization use. At a high level, we can organize the body of existing work in the visualization community based on two categories of stable traits: *personality traits* and *cognitive abilities*. Table 2.1 summaries of the attributes found in the literature.

2.1   COGNITIVE ABILITIES

Cognitive abilities refer to mental capabilities in problem solving and reasoning (including visual reasoning) [80]. The data visualization community has extrapolated the effects of cognitive abilities on the users' performances and experiences with visualizations from foundational research in Psychology. We find literature related to spatial abilities [26, 27, 56, 126, 163, 165, 182], perceptual speed [23, 34, 36, 151, 152, 157, 158], visual working memory [8, 34, 36, 46, 151, 152, 157, 158], and verbal working memory [23, 151, 152, 157, 158]. With a few exceptions (e.g., [36]), high scores in cognitive ability measures usually result in better performances with tasks when using data visualizations.

2.1.1   SPATIAL ABILITY

Another main factor that have been shown to influence visualization use is *spatial ability*. Spatial ability refers to the ability to mentally represent and manipulate two- or three-dimensional representations of objects. Spatial ability is a cognitive ability with a number of measurable dimensions, including spatial orientation, spatial visualization, spatial location memory, targeting, disembedding, and spatial perception [90, 165]. People with higher spatial ability can produce more accurate representations and maintain a reliable model of objects as they move and rotate in space.

There is considerable evidence that these abilities affect how well a person can reason with abstract representations of information, including visualizations. For example, Tversky et al. [161, 162] examined how individual differences in ability affects the extraction of structure and function from diagrams. They showed that participants with high spatial ability create mental models that integrate both structure and function, while participants with low spatial ability form mental models where structure is separate from function. Vicente et al. [166] found that low spatial ability corresponded with poor performance on information retrieval tasks in hierarchical file structures. They found that in general high spatial ability users were two times faster

Table 2.1: Related work with present (✔) and absent (✘) cognitive traits

	Openness	Conscientiousness	Extraversion	Agreeableness	Neuroticism	Locus of Control	Spatial Ability	Perceptual Speed	Working Memory
Allen (1994) [3]	✘	✘	✘	✘	✘	✘	✓	✘	✘
Chen and Czerwinski (1997) [27]	✘	✘	✘	✘	✘	✘	✓	✘	✘
Cohen et al. (2007) [30]	✘	✘	✘	✘	✘	✘	✓	✘	✘
Froese et al. (2013) [56]	✘	✘	✘	✘	✘	✘	✓	✘	✘
Ottley et al. (2015) [127]	✘	✘	✘	✘	✘	✘	✓	✘	✘
VanderPlas and Hofmann (2015) [163]	✘	✘	✘	✘	✘	✘	✓	✘	✘
Velez et al. (2005) [165]	✘	✘	✘	✘	✘	✘	✓	✘	✘
Conati and Maclaren (2008) [36]	✘	✘	✘	✘	✘	✘	✓	✓	✓
Ziemkiewicz and Kosara (2009)[182]	✓	✓	✓	✓	✓	✘	✓	✘	✘
Brown et al. (2014) [21]	✓	✓	✓	✓	✓	✓	✘	✘	✘
Green and Fisher (2010) [66]	✓	✓	✓	✓	✓	✓	✘	✘	✘
Ziemkiewicz et al. (2011) [181]	✓	✓	✓	✓	✓	✓	✘	✘	✘
Ziemkiewicz et al. (2012) [184]	✓	✓	✓	✓	✓	✓	✘	✘	✘
Ottley et al. (2013) [121]	✘	✘	✘	✘	✘	✓	✘	✘	✘
Ottley et al. (2015) [128]	✘	✘	✘	✘	✘	✓	✘	✘	✘
Conati et al. (2014) [34]	✘	✘	✘	✘	✘	✓	✘	✓	✓
Carenini et al. (2014) [23]	✘	✘	✘	✘	✘	✓	✘	✘	✓
Lallé et al. (2017) [97]	✘	✘	✘	✘	✘	✘	✘	✓	✓
Steichen et al. (2013) [151]	✘	✘	✘	✘	✘	✘	✘	✓	✓
Toker et al. (2012) [157]	✘	✘	✘	✘	✘	✘	✘	✓	✓
Toker et al. (2013) [158]	✘	✘	✘	✘	✘	✘	✘	✓	✓

than low spatial ability users and that low spatial ability users were more likely to get lost in the hierarchical file structures.

Chen and Czerwinski [27] found that participants with higher spatial ability employed more efficient visual search strategies and were better able to remember visual structures in an interactive node-link visualization. Velez et al. [165] tested users of a three-dimensional visualization and discovered that speed and accuracy were dependent on several factors of spatial ability. Similarly, Cohen and Hegarty [30] found that users' spatial abilities affects the degree to which interacting with an animated visualization helps when performing a mental rotation task, and that participants with high spatial ability were better able to use a visual representation rather than rely on an internal visualization.

This body of research shows that users with higher spatial ability are frequently more effective at using a variety of visualizations. Taken together, they suggest that high spatial ability often correlates with better performance on tasks that involve either searching through spatially arranged information or making sense of new visual representations. Additionally, there is evidence that high spatial ability makes it easier to switch between different representations of complex information. Ziemkiewicz and Kosara [182] tested users' ability to perform search tasks with hierarachy visualizations when the spatial metaphor implied in the task questions differed from that used by the visualization. Most participants performed poorly when the metaphors conflicted, but those with high spatial ability did not. This confirms findings that spatial ability plays a role in understanding text descriptions of spatial information [42].

2.1.2   PERCEPTUAL SPEED

Researchers have also investigated how perceptual ability influences visualization use. Perceptual speed is a cognitive ability that describes how quickly a person can accurately compare objects. Conati and Maclaren [36] found that a user's perceptual speed influences performance using two difference Geographical Information Systems. They found that users with high perceptual ability performed better with star graphs while users with low perceptual speed performed better with radar graphs.

Allen [3] too found a connection between perceptual speed and system effectiveness. He found that perceptual speed was a significant predictor of spatial scanning ability in search performance. His findings suggest that users often fail to optimize their visualization use for greater search efficiency, and his work calls for the development of user models to automatically guide users toward optimal strategies. Similarly, Conati et al. [34] found that a number of cognitive traits, including perceptual speed, influences performance across different layouts of interactive ValueCharts. Using eye tracking technology, Toker et al. [158] demonstrated how perceptual speed influences interactions with bar graphs and radar graphs. Other work by Toker et al. [157] found that perceptual speed not only impacts performance on visualization tools but also preference for difference tools.

2.1.3   WORKING MEMORY CAPACITY

Many of the studies that investigated perceptual speed also includes measure of visual and verbal working memory capacity. Visual working memory measures the ability to remember the configuration, location, and orientation of an object and verbal working memory describes the ability to remember speech-related information. Prior work has shown that verbal working memory correlates with how people use bar graphs and radar graphs [158]. Results on the impact of visual working memory have been inconclusive.

2.2   PERSONALITY TRAITS

Personality traits are the individual differences in thinking and behaving characteristics [4]. The literature contains numerous examples of personality traits that interacts with visualization use. For instance, researchers have uncovered that *locus of control*, a measure of perceived control, is a key factor that correlates with speed, accuracy and strategy [21, 66, 67, 120, 128, 181, 184]. Prior work (e.g., [21, 66, 67, 184]) shows similar interactions with the dimensions of the *Five-Factor Model*: *extraversion*, *neuroticism*, *openness to experience*, *conscientiousness*, and *agreeableness*.

2.2.1   LOCUS OF CONTROL

LOC [141] measures the degree to which an individual feels in control of, or controlled by external events. Using the Rotter construct [141], individuals are scored on a 23-point scale where the two extreme ends of the scale are categorized as internal LOC and external LOC. Persons who are internally oriented on the scale (*Internals*) believe that external events are contingent upon their own actions while persons who are externally oriented (*Externals*) on the scale believe that events are controlled by powerful beings.

Psychologists have long investigated LOC and research suggests that the differences between Internals and Externals can be quite vast. Internals tend to have a strong sense of self-efficacy allowing them to take control even when faced with difficult problems. Conversely, Externals believe that they have no control over external events, making them far more likely to adapt to situations. However, because of this perceived lack of control, Externals are also more likely to give up when faced with difficulty.

Past research corroborates this. Internality has been shown to correlate with increased effectiveness at work [86], better academic performance [52] and greater ability to cope with stress [7]. LOC also affects learning style. Cassidy and Eachus [24] showed that Internals are more likely to practice deep learning, while Externals are more likely to practice surface learning. This implies that there is a correlation between LOC and general problem-solving techniques which suggests a potential effect of LOC on problem solving using visualizations.

In the medical community, LOC has been shown to affect patients' recovery outcomes. Fisher and Johnstion [53] found that users with external LOC were more likely to become discouraged and give in to their disability. In the visualization community, Green and Fisher [66] showed a significant correlation between LOC and users' speed and accuracy when using hierarchical visualizations. They reported that LOC can be used to predict completion times as well as insight when using the two visualization tools.

2.2.2   BIG 5 PERSONALITY TRAITS

Personality psychology is a well-established area of research, making it a useful lens through which to better understand how different users approach visualization tasks. A common model, the Five-Factor Model, categorizes personality traits on five dimensions: extraversion, neuroticism, openness to experience, conscientiousness, and agreeableness. An individual can be categorized under these personality traits, and longitudinal studies demonstrate that these remain consistent throughout adulthood [140]. Such differences may reflect a user's outlook and common behavior patterns, and recent compelling research on personality differences suggests that they may have a significant effect on performance. Some of these studies indicate that these effects are strongest when tasks are complex, providing a window on how users differ in using a visualization to support higher-level reasoning.

There has been some research in human-computer interaction (HCI) showing that these personality factors are significantly correlated with a user's preference for visual interface designs. For example, a study by Saati et al. on interface skinning in a music player compared preferences for five "skins," or visual themes, that varied only on the dominant color [143]. These results showed that preference for a blue theme was positively correlated with introversion, a yellow theme was preferred by more conscientious users, and more imaginative users preferred the black theme. While user preference may affect adoption rates for a visualization system, differences in performance are more valuable in understanding how people use visualization.

In visualization specifically, prior work has shown that personality traits can significantly affect complex task performance. For example, Ziemkiewicz and Kosara performed an experiment on how conflicting metaphors affect tree visualization evaluations [182]. By varying verbal and visual metaphors in the evaluation conditions, they studied the extent to which different users slowed down in response to metaphor conflicts. The results showed that users who scored highly on the openness to experience dimension of the Five Factor model were unaffected by conflicting verbal and visual metaphors. While this study did not directly compare performance on different types of visual design, it does suggest that participants with high openness may have an easier time switching between different design metaphors, such as those found in a multi-view system.

Green et al. studied the use of visual analytics interfaces by users with varying scores on the Five Factor model [66, 67]. The study investigated search performance across two interfaces for exploring hierarchical genomic information. The authors found that low extraversion and neuroticism scores mapped to more insights, and higher scores on the extraversion and neuroticism scales correlated with faster completion times.

2.3 SUMMARY

An overwhelming body of research has demonstrated how cognitive traits impact people's ability to use information visualization and visualization systems [21, 27, 30, 34–36, 66, 67, 165, 166], and a growing number of researchers have advocated for better understanding of these effects [176, 183]. This prior research demonstrate that the cognitive differences between users may result in dissimilarities in the way information is internalized and thus the way they utilize visualization tools. The experiments in many studies typically include tasks that are commonly performed for data analytics in education and business. Consequently, the experimental discoveries are representative of what could happen in real world scenarios. We will further explore the impact of cognitive traits in Part II.

PART II

Uncovering the Impact of Traits

CHAPTER 3

Spatial Ability and Bayesian Reasoning

"By effectively communicating statistical and probabilistic information, physicians will interpret diagnostic results more adequately, patients will take more informed decisions when choosing medical treatments, and juries will convict criminals and acquit innocent defendants more reliably."

Luana Micallef, Pierre Dragicevic, and Jean-Daniel Fekete   [111]

One advantage of visualization is that it can make complex or abstract concepts easier to grasp by making them visible. In theory, an effective visual representation can make traditionally difficult problems more concrete and easier to understand. In practice, however, nuanced contextual and individual factors heavily impact the effectiveness of visualizations, making the right representation difficult to establish.

In order to build systems that can better facilitate a user's cognitive processes, we must not only understand *which* individual differences can impact reasoning with visualizations but we also need to know *how*. In this chapter, we begin by exploring a real-world decision-making problem to illustrate the practical impact of investigating individual differences. *We demonstrate how problem representation can influence reasoning with medical statistics and how spatial ability mediates Bayesian reasoning.* The findings presented in this chapter have direct implications for medical risk communication.

3.1   INTRODUCTION

As the medical field transitions toward evidence-based and shared decision making, effectively communicating conditional probabilities to patients has emerged as a common challenge. To make informed health decisions, it is essential that patients understand health risk information involving conditional probabilities and Bayesian reasoning [59]. However, understanding such conditional probabilities is challenging for patients [48]. Even more alarming, the burden of communicating complex statistical information to patients is often placed on physicians even though studies have shown they also struggle with accurate estimations themselves [48].

Still, both physicians and patients make life-critical judgments based on conditional probabilities. Deficits in diagnostic test sensitivity and specificity (intrinsic characteristics of the test itself) can lead to false negative and false positive test results which do not reflect the actual state of an individual. For low-prevalence diseases, even a highly specific test leads to false positive results for a majority of test recipients. Unless a patient fully understands the uncertainties of medical tests, news of a negative result can lead to false reassurance that treatment is not necessary, and news of a positive result can bring unjust emotional distress [68].

Misinterpretation of medical test statistics can have serious adverse consequences such as under- and over-diagnosis [112, 172, 173] or even death. However, there are currently no effective tools for mitigating this problem. Despite decades of research, the optimal methods for improving interpretation of diagnostic test results remain elusive, and the available evidence is sparse and conflicting.

Some investigators have attempted to teach Bayesian reasoning skills to physicians and patients, but the reported success rate of these efforts remains low [61]. A promising approach to this problem includes the use of natural frequencies—i.e., countable proportions like *96 out of 1000* instead of *9.6%* [62, 72]. Still, recent research suggests that these techniques have limited utility when used by people without statistical training, with reports that as low as 6% of laypeople solve these problems correctly [111].

Prior work indicates that visualizations may be key for improving performance with Bayesian reasoning problems. For example, results from Brase [20] and work by Garcia-Retamero and Hoffrage [60] suggest that visual aids such as Euler diagrams and icon arrays hold promise. Researchers have also explored visualizations such as decision trees [55, 107], contingency tables [31], "beam cut" diagrams [62], and probability curves [31], and have shown improvements over text-only representations. However, when researchers in the visualization community extended this work to a more diverse sampling of the general population, they found that adding visualizations to existing text representations did not significantly increase accuracy [111, 123, 124].

Given the contradictory findings of prior research, we designed a series of studies that were aim to identify factors that influence performance on Bayesian reasoning tasks. We hypothesized that these discrepancies are due to differences in problem representations (textual or visual), as well as the end users' cognitive ability to reason through these problems when using visualizations. In particular, we proposed that the phrasing of text-only representations can significantly impact comprehension and that this effect is further confounded when text and visualization are incorporated into a single representation. Furthermore, motivated by prior work [27, 88, 165, 166, 182], we also hypothesized that individual differences (i.e., spatial ability) are mediating factors for performance on Bayesian reasoning tasks.

3.2   ESTABLISHING A TEXT-ONLY BASELINE

Many past experiments have used their own Bayesian problems, with differing scenarios, wordings, and framings. For instance, in their work, Gigerenzer and Hoffrage [62] used 15 different Bayesian problems, each with differing real-world implications and potential cognitive biases associated with them (e.g., being career-oriented leads to choosing a course in economics, or carrying heavy books daily relates to a child having bad posture).

In line with our research goals of disambiguating contradictory results in previous research, our first experiment examines how the wording of the text impacts the complexity of Bayesian reasoning problems. In the context of Bayesian problems, the term complexity can have different meanings, for instance: the number of relationships in the problem [88], the number of steps needed to solve the problem, or the amount of information to be integrated or reconstructed [39]. In the current work, we define complexity as the difficulty of extracting information. This hinges on the notion that the simplicity of a task partly depends on the *how* the information is presented. We believe that is it important first to establish a baseline text representation (i.e., **no visualization**) before we consider the effect of adding visualizations.

We conducted an online study and tested three different text-only representations of Bayesian reasoning problems, detailed in Table 3.1. We found that by simply changing how the problem was presented, we observed an improvement in participants' overall accuracy from 5.4–42.4%. We adapted techniques such as probing, which nudges the user to think more thoroughly about the problem, adding a narrative which generalized the problem, and presenting both frames for mitigating framing effects.

This finding gives us insight into how lexical choices of text-only representations of Bayesian reasoning problems govern their effectiveness and may at least partially explain the poor or inconsistent accuracy observed in previous work. By using probing alone, our results showed a significant improvement over our base condition which used direct questioning. This suggests that assessment techniques for Bayesian reasoning problems should be thoroughly scrutinized.

Participants were even more accurate when the stimulus combined all three techniques (probing, narrative, and framing). This finding provides initial evidence that even with text-only representations (i.e., without visualization aids), the phrasing of the problem can impact comprehension. Indeed, there were several factors that potentially contributed to the increase in communicative competence we observed for $Text_{diseaseX}$. For example, using the generic term *Disease X* instead of a specific disease may have mitigated biases introduced by the mammography problem. Alternatively, the observed increase in accuracy could be attributed to the overall readability of the text or the amount of data presented in the conditions (the $Text_{diseaseX}$ condition presented the user with slightly more explicit data than the $Text_{orig}$ and $Text_{probe}$ conditions) (Fig. 3.1). Deciphering these was beyond the scope of this project, but will be an important direction for future work.

Table 3.1: The three questions used in Experiment 1

Text$_{orig}$	10 out of every 1,000 women at age forty who participate in routine screening have breast cancer. 8 of every 10 women with breast cancer will get a positive mammography . 95 out of every 990 women without breast cancer will also get a positive mammography.
	Here is a new representative sample of women at age forty who got a positive mammography in routine screening. How many of these women do you expect to actually have breast cancer? _____ out of _____
Text$_{probe}$	10 out of every 1,000 women at age forty who participate in routine screening have breast cancer. 8 of every 10 women with breast cancer will get a positive mammography . 95 out of every 990 women without breast cancer will also get a positive mammography.
	Imagine 1,000 people are tested for the disease.
	(a) How many people will test positive? _____
	(b) Of those who test positive, how many will actually have the disease? _____
Text$_{diseaseX}$	There is a newly discovered disease, Disease X, which is transmitted by a bacterial infection found in the population. There is a test to detect whether or not a person has the disease, but it is not perfect. Here is some information about the current research on Disease X and efforts to test for the infection that causes it.
	There is a total of 1,000 people in the population. Out of the 1,000 people in the population, 10 people actually have the disease. Out of these 10 people, 8 will receive a positive test result and 2 will receive a negative test result. On the other hand, 990 people do not have the disease (that is, they are perfectly healthy). Out of these 990 people, 95 will receive a positive test result and 895 will receive a negative test result.
	Imagine 1,000 people are tested for the disease.
	(a) How many people will test positive? _____
	(b) Of those who test positive, how many will actually have the disease? _____

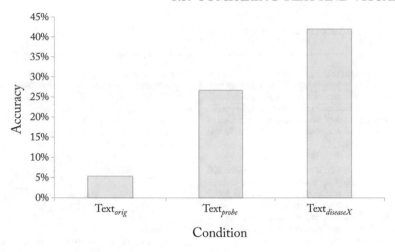

Figure 3.1: Accuracy across all conditions in Experiment 1. Combining probing and narrative techniques proved to be effective for reducing the overall complexity of the text and increasing accuracy.

In the following study, we further address our research goals by investigating the effect of adding visualizations for representing Bayesian reasoning tasks. We use our results from this initial experiment by adopting $Text_{diseaseX}$ as a baseline text-only representation for evaluating different text and visualization designs.

3.3 COMPARING TEXT AND VISUALIZATION

Although visualization has been suggested as a solution to the Bayesian reasoning problem, recent findings suggest that, across several designs, simply adding visualizations to textual Bayesian inference problems yields no significant performance benefit [111, 124]. Micallef et al. [111] also found that removing numbers from the textual representation can improve performance. The findings of this prior work suggest an interference between text and visualization components when they are combined into a single representation.

Differing from prior work which focused mainly on comparing different visualization designs [111], our second experiment aimed to progress Bayesian reasoning research by further investigating the effect of presenting text and visualization together. We examined the amount of information presented to the user and the degree to which the textual and visual information are integrated. Grounded by the baseline condition established in Experiment 1 (Table 3.2 **Control-Text**), we tested representations that gradually integrate affordances of visualizations or the visualization itself.

One affordance of visualizations is that relationships that are implicitly expressed in text are often explicated in visual form. Visualizations make it easier to "see" relationships among

Table 3.2: The six conditions used in Experiment 2

Control-Text

There is a total of 100 people in the population. Out of the 100 people in the population, 6 people actually have the disease. Out of these 6 people, 4 will receive a positive test result and 2 will receive a negative test result. On the other hand, 94 people do not have the disease (i.e., they are perfectly healthy). Out of these 94 people, 16 will receive a positive test result and 78 will receive a negative test result.

Complete-Text

There is a total of 100 people in the population. Out of the 100 people in the population, 6 people actually have the disease. Out of these 6 people, 4 will receive a positive test result and 2 will receive a negative test result. On the other hand, 94 people do not have the disease (i.e., they are perfectly healthy). Out of these 94 people, 16 will receive a positive test result and 78 will receive a negative test result.

Another way to think about this is... Out of the 100 people in the population, 20 people will test positive. Out of these 20 people, 4 will actually have the disease and 16 will not have the disease (i.e., they are perfectly healthy). On the other hand, 80 people will test negative. Out of these 80 people, 2 will actually have the disease and 78 will not have the disease (i.e., they are perfectly healthy).

Structured-Text

There is a total of 100 people in the population.
  Out of the 100 people in the population,
    6 people actually have the disease. Out of these 6 people,
      4 will receive a positive test result and
      2 will receive a negative test result.
    On the other hand, 94 people do not have the disease (i.e., they are perfectly healthy). Out of these 94 people,
      16 will receive a positive test result and
      78 will receive a negative test result.

Another way to think about this is...
  Out of the 100 people in the population,
    20 people will test positive. Out of these 20 people,
      4 will actually have the disease and
      16 will not have the disease (i.e., they are perfectly healthy).
    On the other hand, 80 people will test negative. Out of these 80 people,
      2 will actually have the disease and
      78 will not have the disease (i.e., they are perfectly healthy).

Vis-Only

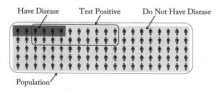

Control+Vis

There is a total of 100 people in the population. Out of the 100 people in the population, 6 people actually have the disease. Out of these 6 people, 4 will receive a positive test result and 2 will receive a negative test result. On the other hand, 94 people do not have the disease (i.e., they are perfectly healthy). Out of these 94 people, 16 will receive a positive test result and 78 will receive a negative test result.

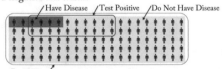

Storyboarding

There is a total of 100 people in the population.

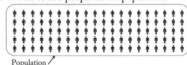

Out of the 100 people in the population, 6 people actually have the disease.

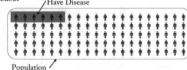

Out of these 6 people, 4 will receive a positive test result and 2 will receive a negative test result.

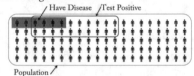

On the other hand, 94 people do not have the disease (i.e., they are perfectly healthy).

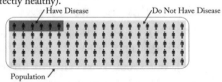

Out of these 94 people, 16 will receive a positive test result and 78 will receive a negative test result.

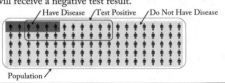

groups. To bridge this information gap, we gradually expanded the text-only representation to explicate implied information and relationships.

Our second experiment also aimed to understand *how* nuances in spatial-ability affect users' capacity to use different representations of Bayesian reasoning problems. Since prior research suggests that low spatial-ability users may experience difficulty when both the text and visual representations are presented [182], we hypothesize that low spatial-ability users would be more adept at using representations which integrated affordances of the visualization but not the visualization itself. On the other hand, we hypothesize that high spatial-ability users will benefit greatly from representations which merge textual and visual forms, as they are more likely to possess the ability to effectively utilize both representations.

The results of our Experiment 1 demonstrated how phrasing of Bayesian problems can influence performance. In Experiment 2, we examined whether we could improve problem representations by enhancing text or combining it with visualization. In an effort to bridge the information gap between text and visual representations, we studied text-only representations that clarified information that usually is more easily seen in a visualization. Our *Complete-Text* design sought to decrease this information gap by enumerating all probability relationships in the text and our *Structured-Text* design used indentations to visualize these relationships. Still, when spatial-ability was not considered, we found that adding more information did not benefit users.

Spatial Ability Matters

We observed similar results with our visualization conditions. Although we hypothesized that the *Vis-Only* design would be more effective than *Control-Text* and *Control+Vis*, our results did not support this hypothesis. Again, when spatial ability was not considered, adding visualizations (with or without textual information) did not improve performance. However, a closer examination of our results adds nuance to this finding when individual differences are considered.

Across all visualizations, spatial ability was a significant indicator of accuracy and completion times. We found that users with low spatial ability generally performed poorly; the accuracy of high spatial-ability users was far higher than the accuracy of low spatial-ability users (78.8% vs. 46.9%). Relative to the *Control-Text* condition, for high spatial users, the *Structured-Text*, *Complete-Text* and *Vis-Only* designs were extremely effective, yielding accuracies of 100%, 90%, and 96%, respectively (Fig. 3.2). These unprecedented accuracies suggest that, for users with high spatial ability, these designs can solve the problem of Bayesian reasoning. However, it is interesting to note that effective designs were "pure" designs (i.e., they did not combine text and visualizations). This finding contradicts prior work in Psychology which demonstrates a multimedia advantage (i.e., providing both text and visual representation) for comprehension and memory [22].

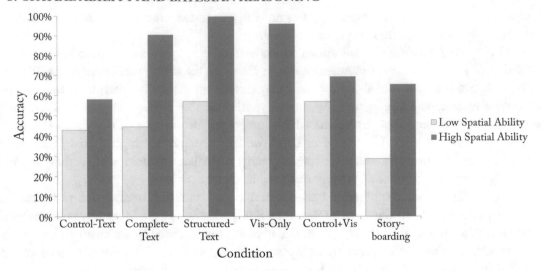

Figure 3.2: Average accuracy for the low and high spatial ability groups for each design. Overall, we found that high spatial users were much more likely to correctly answer the question prompts.

On Combining Text and Visualization

For high spatial-ability users, we found that representations that combined text and visualization (*Control+Vis, Storyboarding*) actually impeded users' understanding of conditional probability when compared to text-only (*Complete-Text, Structured-Text*) or *Vis-Only* conditions. Despite the fact that high spatial-ability users performed comparatively poorly with the *Control+Vis* design (accuracy decreased by nearly 30% when compared to *Complete-Text, Structured-Text*, and *Vis-Only*), such disparity in accuracy was not observed with low spatial-ability users using *Control+Vis*. One possible explanation relies on considering the problem as a mental modeling task. Users with low spatial ability may have simply chosen the representation in *Control+Vis* (text or visualization) that best fit their understanding of the problem. On the contrary, high spatial-ability users may have attempted (and failed) to integrate the text and visualization representations in order to find the correct answer. This hypothesis would be in line with Kellen's [89] hypothesis that text and visual representations in a complex problem may compete for the same mental resources, increasing the likelihood of errors.

The *Storyboarding* design proved to be an enormous obstacle for the user. Performing analysis to investigate the spatial ability scores required to successfully extract information from the six designs revealed that *Storyboarding* demand higher spatial-ability scores than the other designs. While it is intended to gradually guide users through the Bayesian reasoning problem, the different steps may have inadvertently introduced distractors to the information that the user is truly looking for and/or forced users into a linear style of reasoning that was incongruent with their mental model of the problem. This added complexity increased cognitive load to a point that accuracy for all users suffered.

Still, such storytelling techniques have been shown to be effective for communicating real-world data [77, 78, 95, 146, 179]. The tasks in this study, however, go beyond typical information dissemination, as users had to understand information known to be inherently challenging for most people. Future work could investigate the utility of storytelling techniques for similar reasoning tasks.

3.4   SUMMARY

Effectively communicating Bayesian reasoning has been an open challenge for many decades, and existing work is sparse and sometimes contradictory. In this chapter we presented results from two experiments that help explain the factors affecting how text and visual representations contribute to performance on Bayesian problems. With our first experiment, we showed that the wording of text-only representations can significantly impact users' accuracies and may partly be responsible for the poor or inconsistent findings observed by prior work.

Our second experiment examined the effects of spatial ability on Bayesian reasoning tasks and analyzed performance with a variety of text and visualization conditions. We found that spatial ability significantly affected users ability to use different Bayesian reasoning representations. Compared to high spatial-ability users, low spatial-ability users tended to struggle with Bayesian reasoning representations. In fact, high spatial-ability users were almost two times more likely to answer correctly than low spatial-ability users. Additionally, we found that text-only or visualization-only designs were more effective than those which blend text and visualization.

Ultimately, our results not only shed light on how problem representation (both in text phrasing and combining text and visualization) can affect Bayesian reasoning, but also question whether one-size-fits-all visualizations are ideal. Further study is needed to clarify how best to either adapt visualizations or provide customization options to serve users with different needs. The results from these studies can be used for real-world information displays targeted to help people better understand probabilistic information. They also provide a set of benchmark problem framings that can be used for more comparable future evaluations of visualizations for Bayesian reasoning. Further work in this domain can have significant impact on pressing issues in the medical communication field and other domains where probabilistic reasoning is critical.

CHAPTER 4

Big-5, Locus of Control, and Searching Trees

"Locus of Control proved to be an influential personality trait no matter what the interface or task."

Tera Marie Green and Brian Fisher   [66]

Chapter 3 provided empirical evidence of the impact of individual differences on a real-world reasoning task with visualizations. It showed that a user's spatial ability is a mediating factor, and that we can use spatial ability to predict performance across different representations of a Bayesian reasoning problem. Furthermore, we saw that none of the tested conditions were ideal for all users, highlighting the importance of tailoring designs to individuals. In this chapter, we present the results another study that demonstrates the impact of *locus of control* and the *Big 5 personality traits*.

4.1   INTRODUCTION

One of the most consistent and well-studied trait that correlates with performance when using visualizations is LOC [66, 67, 181]. LOC [141] measures the degree to which a person attributes outcomes to themselves (internal LOC) or to outside forces (external LOC). Green and Fisher [66] were the first to uncover a correlation between LOC and performance on two real-world data exploration systems. Participants in their study performed search tasks using two tree visualizations (see Fig. 4.1) and their study revealed three traits that were dominant predictors of speed and accuracy: LOC, neuroticism, and extraversion.

This finding by Green and Fisher demonstrates that performance may be sensitive to nuances of an individual user's cognitive style. When we give users a novel visualization, we are essentially asking them to give up some control over their thinking processes. Some users will find this helpful, while others may find it a hindrance. The designer will therefore benefit from having a sense of how willing a given user will be to take on an external representation, and know how to design a visualization that makes it more or less difficult to ignore the structural

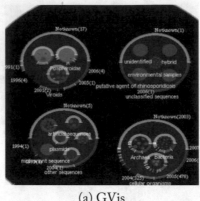

(a) GVis

Vertebrates				(22)
▼ Mammals				(18)
▼ Primates				(5)
Scientific name	Common name	Build	Tools	
Callithrix jacchus	white-tufted-ear marmoset	Build 1.1	⊕ Ⓑ Ⓡ	
Homo sapiens	human	Build 37.2	⊕ Ⓑ Ⓡ Ⓖ Ⓖ	
		Build 36.3	⊕ Ⓑ Ⓡ Ⓖ	
Macaca mulatta	rhesus macaque	Build 1.2	⊕ Ⓑ Ⓡ Ⓖ	
Pan troglodytes	chimpanzee	Build 2.1	⊕ Ⓑ Ⓡ Ⓖ	
Pongo abelii	Sumatran orangutan	Build 1.2	⊕ Ⓑ Ⓡ	
▼ Rodents				(2)
Scientific name	Common name	Build	Tools	
Mus musculus	laboratory mouse	Build 37.2	⊕ Ⓑ Ⓡ Ⓖ Ⓖ	
		Build 36.1	⊕ Ⓑ Ⓡ	
Rattus norvegicus	rat	RGSC v3.4	⊕ Ⓑ Ⓡ Ⓖ Ⓖ	
▶ Monotremes				(1)
▶ Marsupials				(1)
▶ Other Mammals				(9)
▶ Other Vertebrates				(4)
▶ Invertebrates				(14)
▶ Protozoa Ⓑ				(18)
▶ Plants ⊕				(46)

(b) NCBI Map Viewer

Figure 4.1: The two interfaces used in Green and Fisher's study [66] of personality differences in visual analytics use.

aspects of that representation. The practitioner will also benefit form having a set of concrete guidelines that maps user characteristics to variable features of the system.

The two systems explored by Green and Fisher differed on many dimensions, including the use of color, labeling, interaction, and layout style, therefore making it difficult to generalize the findings. We believed that layout style is the key variable that determines the interaction between LOC and compatibility with different system designs. The definition of layout here encompasses any differences in the spatial arrangement and presentation of marks in a visualization. This is to be distinguished from differences in the visual encoding, that is, how individual data variables are mapped to individual graphical variables, such as color or size.

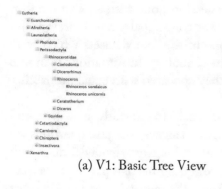

(a) V1: Basic Tree View

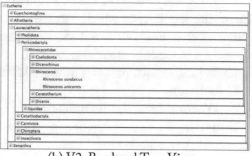

(b) V2: Bordered Tree View

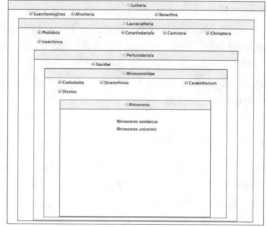

(c) V3: Indented Boxes View

(d) V4: Nested Boxes View

Figure 4.2: The four visualizations used in the study. Each view is showing the same portion of one of the phylogenetic tree datasets.

4.2   EXPERIMENT

To test our hypothesis, we conducted an online study with subjects who varied in their LOC. Participants are presented with four variations of a hierarchy visualization showing phylogenetic data (Fig. 4.2). These designs include a view that employs a list-like organizational structure (V1), a view that presents the hierarchy in a strong containment metaphor (V4), and two designs that lie between these extremes (V2 and V3). Our main hypothesis was that people with a more external locus of control are more willing to adapt their thinking to unfamiliar visual metaphors than those with an internal locus of control. Specifically, an individual with a more internal LOC will show a performance decrease when using layouts with a strong containment metaphor, while those with a more external locus of control will not show this decrease.

The main portion of the experiment consisted of four sessions, one with each of the four views. The sequence in which participants saw these views was counterbalanced to prevent ordering effects. Each view was randomly associated with one of the four datasets which were subset of a taxonomic tree from the National Center for Biotechnology Information's Genome database [115]. Each participant saw all four datasets, but they appeared with equal probability in all four views.

We considered the unfamiliarity of the datasets to be beneficial to our study, since we could trust that participants would need to consult the views in order to answer the task questions we presented them. Following Green and Fisher, [66], these questions were divided into search tasks and inferential tasks. In both cases, tasks took the form of questions that participants were expected to consult the visualization to answer. This is similar to the methodology used in most visualization evaluation studies. These two question types are meant to represent simple data lookup and more complex analytical tasks, although they are simplified versions of the real-world versions of these tasks. We expect to see more differences in the inferential questions, since these are more likely to require understanding of the structure of a dataset rather than simple navigation ability.

4.3   RESULTS

Our task questions proved to be quite difficult, with an overall accuracy of 68.6% correct responses on search tasks and 47.1% on inferential tasks. This difficulty, and the large amount of time spent interacting with the views, should be kept in mind when interpreting the following results. Across all participants and question types, no view condition was more or less difficult in terms of accuracy or correct response time. As our primary interest is in how these results varied with a participant's personality scores, further analysis focuses on participants grouped by personality type. Generally, we found support for our hypothesis that participants with a more internal LOC would have more difficulty with views more similar to V4. While we found that participants with a more external LOC did perform very well with V4, we did not find a corresponding trend in which they performed more poorly on views similar to V1.

4.3.1   LOCUS OF CONTROL

We divided participants into three groups based on their score on the LOC scale. Participants with a score lower than one standard deviation from the mean (i.e., less than 3.01) were classified as *external LOC* users. Those with a score greater than one standard deviation from the mean (i.e., greater than 4.21) were classified as *internal LOC* users. The rest were classified as *average LOC* users.

Our results replicated those of Green and Fisher [66] in some cases but not completely. Error rates across the two experimental designs are not directly comparable, as in their case, participants were allowed to try as many times as needed to answer a question correctly, with each mistake recorded as an error. However, in both our study and theirs, we found that participants

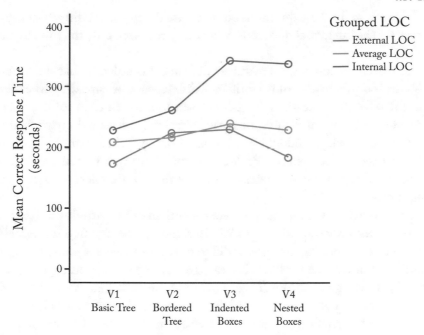

Figure 4.3: Response time for correct answers only across the four view conditions and three Locus of Control (LOC) groups. Participants with a highly internal LOC, who see themselves as in control of external events, were much slower than other participants at answering question in V4, a visualization that uses a strong nested-boxes visual metaphor. Participants with a highly external LOC, who see themselves as controlled by outside events, are relatively more likely to perform quickly on V4.

made more errors overall in the containment-metaphor visualization than in the more list-like view (V1 in our case, Map Viewer in theirs). More significantly, in both our work and theirs, participants with an external LOC responded faster to inferential questions than did participants with an internal LOC, particularly in a visualization with a containment-based visual metaphor (V4 in our case, GVis in theirs).

Together, these two sets of findings provide evidence that the effect of LOC on visualization use is a robust one. Users with either an internal or external LOC show performance differences in general on data exploration tasks, and additionally, each group performs better with different visualization styles. This suggests that LOC is a variable that merits further study, and that personality differences are a valuable topic of research in visualization.

In general, we did not replicate their findings on search question response times. They found that internal LOC participants responded faster to search questions in GVis. However, their completion times for these tasks included any incorrect responses and subsequent guesses made by participants, so for these questions our response times are not directly comparable. This was not the case for the inferential questions, where they recorded only the time to make a single response.

Since our search questions were relatively difficult, it is unlikely that the lack of an effect in these questions is only a matter of performance differences not appearing in easier tasks. We speculate that the inferential questions forced the users to think in terms of the structure of the data to a greater degree. The search questions may have simply measured a participant's ability to navigate the interface quickly, while the inferential questions asked them to characterize parts of the data in an open-ended fashion. Participants may have interpreted these questions in a variety of ways, allowing the structural elements of the visualization design to play a greater role in their thought process.

Although external LOC participants were faster than other participant groups at answering questions in V4, they were equally fast in V1. This finding does not fit our original hypothesis that internal and external participants prefer different types of visual layouts. It may be that the very high familiarity of tree menus like V1 created a training effect that caused it to break the overall pattern. However, given the evidence, we cannot conclude that external participants perform better with containment views than with list-like views. Rather than a clear trend of group preference, a better interpretation of our results may be that external LOC participants are generally better able to answer inferential types of questions using unusual visualization layouts. An experiment that removes the potential confound of a highly familiar view would be needed to test which interpretation is better supported.

Familiarity may also explain the higher preference scores across all participants for V1 and V2, although it is interesting that there was no correspondence between preference and performance. It is also possible that this lack of a relationship may reflect the fact that our participants were paid a bonus for correct responses, and therefore had an incentive to perform well despite disliking the interface. In any case, people may have felt that V3 and V4 were especially confusing due to their unusual appearance, but they were just as capable of answering questions with these interfaces.

4.3.2 BIG 5 PERSONALITY TRAITS

Compared to Green and Fisher [66], we found fewer notable differences between participants categorized by neuroticism or extraversion. In fact, these personality dimensions had no significant effect on response time, which is where the most dramatic effects of LOC were found. On the other hand, we did find that these dimensions influenced participants' accuracy on search tasks.

In the case of neuroticism, our results provide support for Green and Fisher's finding that more neurotic participants generally perform better on search tasks with visual interfaces. The highly neurotic participants were significantly more accurate overall, aggregating all four views and both question types. As suggested by Green and Fisher, this may be explained by the theory that people with more neurotic or trait-anxious personalities tend to be more attentive to problem-solving tasks up to a certain level of complexity [79].

However, we also found that this effect was especially pronounced in the case of V3 and V4, the most container-like of the four views. Notably, the low-neuroticism participants performed more poorly on these two views than on V1 and V2, which likely contributes to the overall relationship between neuroticism and accuracy (Fig. 4.5). This mirrors the findings on LOC (see Figs. 4.4 and 4.5) and suggests that less neurotic participants, like the more internal ones, are less able to make sense of these types of visual layouts. It is possible that the greater attentiveness of these users makes it easier for them to learn an unfamiliar interface. Lower neuroticism also correlates with a more internal LOC, and it is possible that the two scales measure similar qualities that both indicate different aspects of a user's unwillingness or inability to adapt to unusual external representations. An alternate explanation is that the neurotic participants may put more pressure on themselves to complete a question correctly rather than abandoning a task due to an unfamiliar visualization.

In the case of extraversion, our results seemingly diverge from Green and Fisher's. They found that more extraverted participants responded more quickly to search tasks, and we found that more introverted participants were more accurate on all task types. One possibility is that,

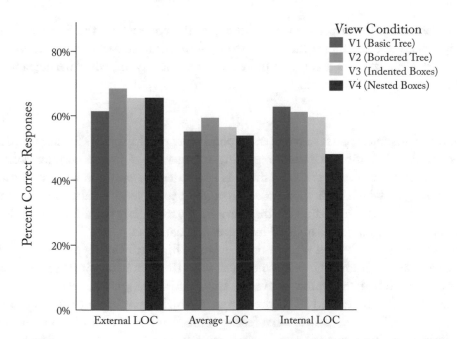

Figure 4.4: Percentage of correct answers across the four view conditions for participants grouped by their Locus of Control score. Participants with a more external LOC were more accurate overall, while the other groups performed poorly with V4.

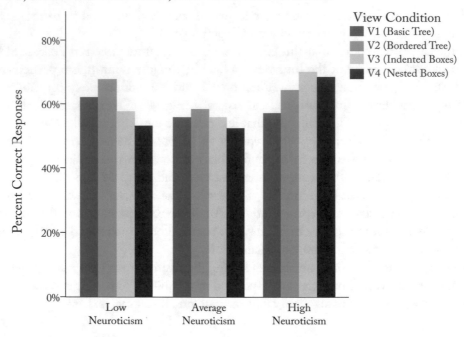

Figure 4.5: Percentage of correct answers in both question types across the four view conditions for participants grouped by their neuroticism score. More neurotic participants performed more accurately with visualizations that used a more container-like layout, while participants in other groups displayed the opposite trend.

under some circumstances, extraverted users respond more quickly but less accurately. That said, we did not find any significant results regarding response time in the current work, so this remains speculative. However, this hypothesis is supported by previous work on introversion and problem solving which finds that more introverted people tend to take more time to think through problems [110]. This extra time may have been particularly helpful in a situation where they had to reason with unfamiliar visual interfaces and data sets.

Since extraversion showed no significant relationship with view type, the overall profile of a participant in our study who performed well with the more containment-like visual layouts is someone who had an external or average LOC and was highly neurotic. As neuroticism measures emotional stability, and LOC the degree to which a participant feels in command of situations, these findings may indicate that feeling somewhat out of control can be an advantage when it comes to making sense of novel visualization designs. It is worth noting that both of these dimensions tend to be negatively correlated with job performance and other practical outcomes [86]. Taken together, this suggests the possibility that persons who struggle with more standard tasks are better suited to thinking with complex visual representations. This possibility warrants further investigation. At the same time, a novel visualization design, while helpful for some users, may be a hindrance to users who already perform a task well using their own methods.

Further work is needed to understand these patterns of users who perform better or more poorly as visual layouts tend toward a strong containment metaphor. It is necessary to examine whether these findings can be generalized beyond this specific task and data type as well as whether they can be generalized to different sets of visual metaphors. Nonetheless, we argue that our findings, particularly those on LOC, serve as a step toward a better understanding of the externalization theory of visualization and how it works for different types of users.

4.4   SUMMARY

This chapter demonstrates how users with different personality types react to varying layout styles used in a hierarchy visualization. We found evidence that systematic differences in layout style can indeed influence a user's response time and accuracy with different types of visualizations that are informationally equivalent but differ in layout. These findings seem to fit a pattern in which users with a more external LOC are more efficient at using a visualization which uses a highly explicit visual metaphor than users with a more internal LOC. We hope that these findings can serve as a step toward better understanding of why subtle differences between users' personality styles can have a surprising influence on visualization use.

CHAPTER 5

Locus of Control and Strategy

"Many of the guidelines that inform how designers create data visualizations originate in studies that unintentionally exclude populations that are most likely to be among the *'data poor'*."

Evan M. Peck, Sofia E. Ayuso, and Omar El-Etr   [66]

The work in Chapter 4 we saw that LOC impacts both speed and accuracy. Is it possible to customize or design an interface based on LOC? Such customization would hinge on the assumption that LOC not only impacts speed and accuracy, but it also influences *how* a user interacts with an interface. In this chapter, we explore the feasibility of personalized visualizations by analyzing users' interactions to investigate the relationship between LOC, visual design, and strategies.

5.1   INTRODUCTION

Interactions are critical to the success of visual analytic tools, and interactive interfaces as a whole [156]. It is through interaction that humans leverage their curiosity, intuition, and creativity to discover patterns, relationships, and other phenomena within data. Interaction fosters visual data exploration. Additionally, users can test and assert hypotheses through their interactions [133, 156]. Thus, gaining a thorough scientific understanding of how users can benefit from interactions in visualization is an important research area [133].

The ability to interact in order to explore information has important implications with regards to enabling the cognitive processes involved in gaining insight into data. Such processes, which can be broadly categorized as sensemaking tasks [91, 134], involve a series of cognitive manipulations and transformations of the data. These tasks fundamentally involve transferring the domain expertise and hypotheses from a user to a system through user interaction. Thus, designers of visual analytic tools often aim to create user interactions that enable and support this cognitive flow of the analytic process [50]. The challenge then is understanding to what extent the interactions a user performs reflects their strategies, mental models, and analytic processes. Further, how much of these cognitive artifacts can be recovered through the analysis of interaction data?

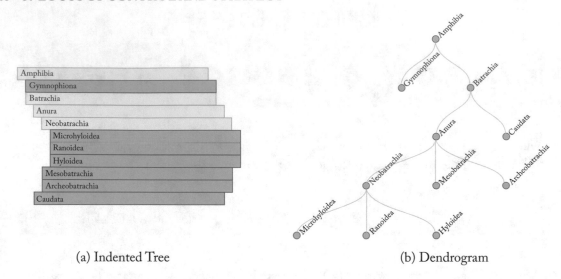

(a) Indented Tree   (b) Dendrogram

Figure 5.1: The two visualizations used for our experiment.

We hypothesize that the observed differences in speed and accuracy of LOC groups are due to differences in their strategies. We also believed we could infer strategies and cognitive traits from their interaction logs. By better understanding how individuals use visualizations, the community can begin to design tools that target users' specific cognitive needs. We focus on how LOC impacts users' search strategies—their data exploration path as they perform searches using a given visualization.

5.2   EXPERIMENT DESIGN

We designed an experiment to examine exactly how LOC impacts strategies. We recruited 54 participants over Amazon's Mechanical Turk service (28 males), and we used the LOC inventory from the International Personality Inventory Pool [63]. Subjects in the study completed search tasks using two hierarchical visualizations: an indented tree (Fig. 5.1a) and a dendrogram (Fig. 5.1b). These particular visualizations were chosen because they are typical representations of hierarchical data [148], are commonly used in real world scenarios and have been extensively studied in the visualization community [66, 121, 185]. The indented tree uses indentation to depict hierarchy while the dendrogram uses a classic node-link structure.

With the exception of the layout, all other the design features were consistent across the two visualizations. Participants were able to explore the datasets by clicking parent nodes to expand their children. Clicking an already expanded node "hides" that subtree. If a user clicks a parent node to expand a "hidden" subtree, the subtree would be restored to its former state. We used two different datasets; both were subsets of a full taxonomic tree retrieved from National

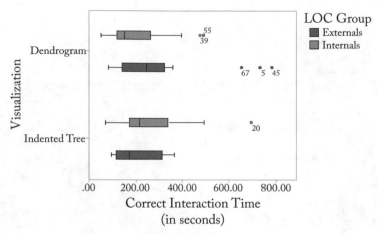

Figure 5.2: Correct response time for the two LOC groups across the two visualization designs.

Center of Biotechnology Information's Genome database [115]. Each dataset was a phylogeny tree where the leaf nodes were actual species. By hovering over a leaf node, participants were able to access attributes of the species.

5.3   RESULTS

Each participant performed two search tasks (one with each visualization), resulting in a total of 108 trials. To filter the inherent "noise" in our Mechanical Turk data, our analysis only includes trials where the participant successfully found the target (91 trials). The average time spent completing the task was 259 sec ($\sigma = 207.7$) with an average of 75 clicks ($\sigma = 11$), and the overall accuracy rate was 68% (75.6% with the indented tree and 62% with the dendrogram). While participants were slightly more accurate with the indented tree, the observed difference was not statistically significant.

5.3.1   INTERACTION TIME

We divided participants into two groups based on the median split of their LOC scores. Figure 5.2 summarizes our interaction time findings. Taking a closer look at interaction times, we found that Externals were almost two times faster with the indented tree visualization (Fig. 5.1a) than the Internals. The Externals were also almost two times faster with the indented tree than they were using the dendrogram (Fig. 5.1b). Interestingly, we observed the exact opposite effect with the Internals.

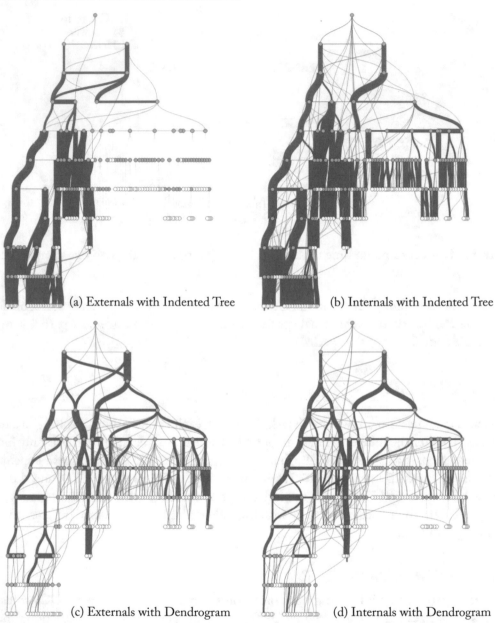

(a) Externals with Indented Tree

(b) Internals with Indented Tree

(c) Externals with Dendrogram

(d) Internals with Dendrogram

Figure 5.3: The figures above show the aggregated exploration paths for users who were successful in locating the target. We grouped users based on a median split of their LOC scores and the visualization design used. For ease of comparison, we used a dendrogram to visualize the exploration paths of both visual designs used in our experiment. The weight of the links represent the percentage of users from that group who traversed that link during their search.

5.3.2   STRATEGIES

During the experiment, a user explored a non-leaf node by clicking to expand or collapse its children and explored a leaf node by hovering to view its attributes. For our analysis we recorded each participant's mouse click and mouse hover events.

With this data, we were then able to reconstruct each user's exploration path as they searched for the target species. This analysis was performed manually using a visualization tool that we developed specifically for this study, and utilizes animation and juxtaposition to compare different exploration paths. Figure 5.3 summarizes our findings.

In general, we found that visualization design influences a user's strategies. When using the indented tree visualization, most participants explored the tree in a top-down fashion, resulting in a search strategy that resembles a depth-first search. Search patterns with the dendrogram were usually less structured and were often a combination of depth-first and breadth-first search.

Again, we separated participants into LOC groups. When using the indented tree visualization, Externals tended to be very strategic and explored the tree in top-down fashion, following the search strategy afforded by the visualization design (Fig. 5.3a). While this strategy logically follows the indented tree's top-down design, we found that Internals were far less likely to adopt this strategy (Fig. 5.3b). This resulted in a more exploratory but less effective search strategy.

Conversely, when using the dendrogram visualization, we found that Externals adopted more sporadic search techniques, resulting in much slower interaction times (Fig. 5.3c). On the other hand, Internals performed a combined depth- and breadth-first search that proved quite effective with the dendrogram (Fig. 5.3d).

5.4   DISCUSSION

While we seldom think about a user's personality traits when designing interfaces, the current work demonstrates why they must be considered. Between the tested designs, no design was suitable for all users. We found that Externals performed better with the indented tree but the Internals were more effective with the dendrogram. When asked which visualization they preferred using, 78% of Externals reported to prefer the indented tree over the dendrogram and 63% of Internals said they preferred the dendrogram.

These results are consistent with the LOC construct. Internals prefer control and tend to struggle when a visualization does not fit the their mental model of a problem [185]. Therefore, a mismatch in design can result in a gulf of evaluation or execution [116]. *The gulf of execution* describes the difference between the user's intentions and how well the design supports their goals. *The gulf of evaluation* is the difference between the system's state and the user's perceived state of the system.

While both visualizations were fully exploratory, the design of the indented tree best facilitates a top-down exploration and users who adapted this strategy were fastest (see Fig. 5.3a). This design property is subtle, but it proved to be a hindrance for the Internals. Conversely, since Externals in general do not believe that they can influence external events, they are more adaptable. The gulfs were small for Externals, and they were able to adopt the top-down search strategy afforded by the indented tree.

The scenario was reversed for the dendrogram visualization. The dendrogram proved to be more versatile and as a result, the Internals excelled with this visualization. They were able to freely search in a manner that best fit their mental model of the data. Unfortunately, this versatility was incompatible with the Externals. Their perceived lack of control coupled with the lack of guidance may have been too overwhelming.

Our design implication follows directly from this finding. When designing for Internals, designers should ensure that designs are flexible to users' mental models. Internals perform better when a visualization allows them to explore the data freely and doesn't impose a strategy. However, the needs of Externals are quite different. While they are in general more flexible and are likely to adapt to novel designs, our findings suggest that adapting to a flexible system may be overwhelming. Unlike Internals who take control of difficult situations, Externals are more likely to feel hopeless and may give up. Thus, it is essential to provide guidance (implicit or explicit) to Externals.

That said, LOC is a multidimensional construct and there are also many other cognitive traits and states that may impact performance on visualization systems [130]. Thus, it is important for us to better understand how these factors, individually and collectively, impact the usability of our designs. There is also evidence that certain traits can be detected from users' interactions [21, 151]. Future work can expand this existing research by exploring how systems can automatically detect and adapt to an individual's specific needs.

5.5   SUMMARY

Altogether, we have demonstrated that individual differences, specifically the personality trait LOC, do impact both interaction times and search strategies. We believe that our work is a significant step toward fully understanding *how* individual differences affect visualization use and how we can begin to design visualizations that better facilitate users' cognitive needs.

PART III

Toward Adaptation

CHAPTER 6

Learning from Interactions

"It is important that the research community's focus on better understanding the relationship between inquiry and interaction not lose sight of the fact that analytic interaction is embedded in a user's experience in the world."

William A. Pike, John Stasko, Remco Chang, and Theresa A. O'Connell   [133]

As we saw in Chapter 5, interactions with a visualization can reflect important high level information about users, such as their strategies and reasoning processes. However, such manual analysis is limited in terms of scale and timeliness. In order to truly leverage the information contained in interaction logs, automated methods must be developed and evaluated based on their ability to infer high level information about users. The work in this chapter demonstrates on a small visual analytics task that it is possible to automatically extract high-level semantic information about users and their analysis processes. Specifically, by using well-known machine learning techniques, this chapter shows that we can: (1) predict a user's task performance and (2) infer personality traits. Furthermore, we establish that these results can be achieved quickly enough that they could be applied to real-time systems.

6.1   INTRODUCTION

Visual analytics systems integrate the ability of humans to intuit and reason with the analytical power of computers [87]. At its core, visual analytics is a collaboration between the human and the computer [40]. Together, the two complement each other to produce a powerful tool for solving a wide range of challenging and ill-defined problems.

Since visual analytics fundamentally requires the close collaboration of human and computer, enabling communication between the two is critical for building useful systems [156]. While the computer can communicate large amounts of information on screen via visualization, the human's input to an analytic computer system is still largely limited to mouse and keyboard [100]. This human-to-computer connection provides limited bandwidth [82] and no means for the human to express analytical needs and intentions, other than to explicitly request the computer to perform specific operations.

Researchers have demonstrated that although the mouse and keyboard appear to be limiting, a great deal of a user's analytical intent and strategy, reasoning processes, and even personal

Figure 6.1: Screenshot of the interface used in the study. Users perform the Where's Waldo [70] search task by finding Waldo, the man drawn on the right, in the image. The navigation commands are simple: zoom in or out, pan up, down, left, or right.

identity can be recovered from this interaction data. Machine learning researchers have recovered identity for re-authenticating specific users in real time using statistics over raw mouse interactions [109, 137, 138, 178] and keyboard inputs [99], but classified only identity, no user traits or strategies. In visual analytics, Dou et al. [47] have shown that strategies can be extracted from interaction logs alone, but at the cost of many hours of tedious labor. Unfortunately, these manual methods are not feasible for real-time systems to adapt to users. The techniques needed to learn about users and their strategies and traits in real time do not exist to our knowledge.

We conducted an online experiment to simulate a challenging visual search task that one might encounter as a component of a visual analytics application with the game *Where's Waldo* (see Fig. 6.1). In the main task, participants navigated a Where's Waldo image by clicking the interface's control bar (Fig. 6.1). The control bar was designed to resemble Google Maps' interface and afforded six interactions: *zoom in*, *zoom out*, *pan left*, *pan right*, *pan up*, and *pan down*. However, unlike Google Maps, our interface does not allow dragging, rather all actions occur through mouse clicks only.

The zoom levels for the interface range from 1–7 (level 1 being no zoom and level 7 being the highest magnification possible). The full image has resolution 5646 by 3607 pixels. At zoom level 1, the full image is shown. At zoom level k, the user sees proportion $1/k$ of the image. Panning moves the display by increments of $1/2k$ pixels.

The interface also includes two buttons not used for navigation: *Found* and *Quit*. When the target is found, the participant is instructed to first click *Found* then click on the target. The user must then confirm the submission on a pop-up alert. We require multiple clicks to indicate Waldo has been found to encourage participants to actively search for the target instead of repeatedly testing many random guesses. If the participant clicks *Found* but does not click on the correct location of Waldo, the click is logged, but nothing happens visually. Unless the participant quits the application, the experiment does not terminate until Waldo is found correctly.

For our analysis, we recorded as much mouse activity as possible, including both mouse click and mouse move events. Mouse click events on interface buttons were logged with a record of the specific button pressed and a time stamp. Similarly, we recorded the interface coordinates of the mouse cursor and the timestamp for every mouse move event. To establish labels for our machine learning analysis of performance outcomes and personality traits, we recorded both completion time and personality survey scores for each participant.

We collected data at the lowest possible level to ensure that we captured as much information about the participants' analysis process as possible. Over the next four sections we discuss how we first visualize this data, then create encodings to capture different aspects of the participants' interactions based on three core aspects of visual analytics: data, user, and interface. Specifically, we encode (1) the portion of the data being displayed, as high-level changes in program state, (2) low-level user interactions, in the form of complete mouse-event data, and (3) interface-level interactions, as sequences of button clicks on the interface's controls. We analyze our data with these encodings with machine learning to evaluate the following hypotheses. First, we hypothesize that participants who are quick at completing the task employ different search strategies from those who are slow, and that these differences are encoded in a recoverable way in the interactions; second, that we can analytically differentiate users' personality traits based on interactions; and third, that these differentiations can be detected without collecting data for the entire timespan of the task, but instead can be found using a fraction of the observation time.

6.2   VISUALIZING USER INTERACTIONS

To explore our hypothesis that we can detect strategies employed by different groups of participants, we first visualize their interactions. Figures 6.2 and 6.3 show example visualizations of user movement around the Waldo image. The area of the visualization maps to the Waldo image. Each elbow-shaped line segment represents a transition from one user view of the image to another, i.e., from a view centered on one end-point of the line to the other. Where these lines intersect with common end-points are viewpoints of the image experienced by the participant while panning and zooming. The lines are bowed (elbow shaped) to show the direction of movement from one viewpoint to the next. Lines curving below their endpoints indicate movement toward the left, and those curving above indicate movement to the right. Bowing to the right of the viewpoints indicates movement toward the bottom, and bowing left indicates movement toward the top.

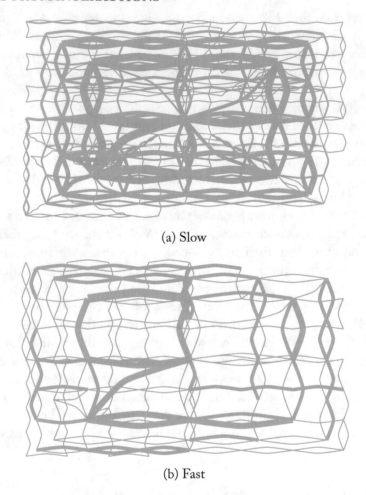

(a) Slow

(b) Fast

Figure 6.2: Visualizations of transitions between viewpoints seen by participants during the study (see Section 6.2). Subfigures (a) and (b) show slow and fast users, respectively, as determined by the mean_nomed splitting method (see Section 6.3).

Zoom levels of viewpoints are not explicitly encoded, but the set of possible center points is determined by the zoom level. High zoom levels mean center points are closer together, so shorter-length lines in the visualization indicate the user was exploring while zoomed in. Note that diagonal movement through the Waldo image is not possible directly with the participants' controls. Instead, diagonal lines in the visualization are created because of zooming, i.e., when zooming out requires a shift in the center point.

This visualization can be used to show the movements made by a whole group of users by counting, for each flow line, the number of users who made the transition between the two

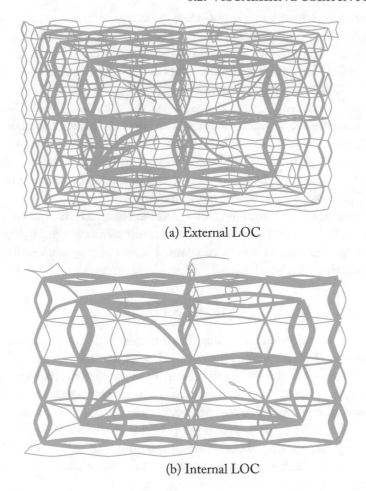

(a) External LOC

(b) Internal LOC

Figure 6.3: Visualizations of transitions between viewpoints seen by participants during the study (see Section 6.2). Subfigures (a) and (b) are split with the mean_nomed method (see Section 6.3) based on LOC, a personality measure of a person's perceived control over external events on a scale from externally controlled to internally controlled.

corresponding viewpoints in the correct direction. In our examples, we are showing such aggregations for four different groups of users. In each case, the thickness of the lines encodes how many users in the group made that transition.

The two sub-figures of Fig. 6.2 compare users who were fast vs. slow at completing the task. Users were considered fast if their completion time was more than one standard deviation lower than the mean completion time, and correspondingly considered slow with completion times more than one standard deviation above the mean (for further explanation see Section 6.3). Users who were slow produce a finer-grain set of lines, indicating they made more small movements through the image using a higher zoom level and saw more of the Waldo image in higher detail. Further, the extra lines in the lower left of Fig. 6.2a as compared to Fig. 6.2b suggest that these slower participants were led astray by the distractors in the image, e.g., the people wearing similar clothing to Waldo seen in Fig. 6.1, insets (b) and (c).

Evidence of different strategies is also salient when visualizing results based on some personality factors. The personality trait LOC has been shown to affect interaction with visualization systems [66, 121, 185]. Figures 6.3a,b visualize differences between participants with external (low) vs. internal (high) LOC. In these subfigures, we see that the external group zoomed in much further on average, while the internal group performed more like the fast group and was able to find Waldo with a smaller set of viewpoints.

These observations are readily seen through these visualizations, but cannot be seen from inspection of the data, nor from machine learning results. Encouragingly, these visualizations hint that there are patterns to uncover in the data. The rest of this work explains our analytical results in extracting them automatically with machine learning.

6.3   COMPLETION TIME FINDINGS

In Section 6.2, we presented visual evidence that our collected interaction data encodes differences between groups of participants. However, being able to tell fast users from slow is more useful if it can be done automatically. In this section, we delve into the data with analytical methods, using machine learning to build predictors of task performance outcomes. In particular, we adopt two common machine learning algorithms, decision trees [113], which learn hierarchical sets of rules for differentiating data, and support vector machines (SVMs) [74], which learn hyperplanes that separate data points of different classes in the data space. We apply these, to three representations of the interaction data, created to capture different aspects of how users interacted with the system.

Specifically, we tested three categories of representations of the participants' interactions, corresponding to some core aspects of visual analytics (data, user, and interface): the views of the image data participants encountered during their task (state-based), their low-level mouse events (event-based), and their clicks on interface controls (sequence-based). In this section we briefly explain how we derive the target participant groups used for our machine learning results, then show, for each representation of the data, our results at predicting if a given user would be fast or slow in completing the task.

We establish two different methods for labeling our participants based on the collected data. Our analyses aim to classify participants into discrete classes, fast and slow, but our recorded data includes only each participant's actual completion time. The first discretization method is to apply the mean completion time (469.5 sec) as a splitting point: participants with a completion time lower than the mean are assigned to the "fast" group, and higher to "slow." Participants with scores exactly equal to the mean are excluded from the data. In our results, this splitting method is indicated as *mean*. In the second method, we assume that participants whose scores are within one standard deviation of the mean have "average" performance and we exclude them from the study, labeling the rest as above. We refer to this approach as the "no-medium" splitting method, indicated in results tables as *mean_nomed*. The no-medium method allows us to see that stronger patterns emerge for participants with more extreme performance.

6.3.1   STATE-BASED ANALYSIS

In the visualization of participants' movement through the Waldo image (see Section 6.2), differences across groups of participants in how they examine the data become salient. This discovery would be more broadly applicable if the differences could be determined automatically. We create two data representations emulating these visual forms to search for patterns that differentiate users based on what parts of the image they chose to look at. In the "state space" encoding, we capture the portion of the data viewed as each participant navigated the Waldo image. In the "edge space" encoding, we capture transitions participants made between viewpoints of the image. Applying SVMs yields high-accuracy classifiers of completion time with both representations.

The state space encoding can be represented by a vector space. We consider the set $s \in S$ of all visual states (given by view position and zoom) that were observed by any user during completing the task. We create a set of vectors u_i, one representing each user, such that $u_i = (count_i(s_1), count_i(s_2), \ldots, count_i(s_{|S|}))$, where $count_i(s_j)$ indicates the number of times user i landed on state j. For the data from the Waldo task, this process yields a vector space in 364 dimensions.

A similar vector space expresses the transitions between viewpoints of the visualization, encoding how participants moved the viewpoint around the image in their search for Waldo. Their strategies may be encapsulated by how they directed the view during their search. In this vector space, the set $t \in T$ consists of all transitions made between any viewpoints by any participant while completing the task. If each viewpoint is represented by the location of its center, x, then $T = \{(k, m)\}$ where any participant made the transition $x_k \rightarrow x_m$ from position x_k to position x_m while searching for Waldo. Each individual user's vector is constructed as $v_i = (count_i(t_1), count_i(t_2), \ldots, count_i(t_{|T|}))$, where $count_i(t_j)$ indicates the number of times user i made transition t_j. The dimensionality of our derived transition-based vector space (edge space) is 1134. The zoom levels are not explicitly encoded, but the set of possible center points is determined by the zoom level. This feature space is most closely related to the visualization described in Section 6.2 and seen in Fig. 6.2.

The calculated vectors are used as a set of data features for input to an SVM [164], a widely applied machine learning method that works on vector space data. SVMs are both powerful and generic, and work by discovering an optimal hyperplane to separate the data by class. For this work we focus on results from the default implementation in the machine learning software package Weka [69], which means a linear hyperplane, and slack parameter $c = 1$. This choice

Table 6.1: Completion Time Classifiers—results for state space, edge space, and mouse events were achieved using support vector machines. The n-gram space results use decision trees. These results were calculated using leave-one-out cross validation.

Data Representation	Class Split	Accuracy (%)
State space	Mean_nomed	83
	mean	79
Edge space	Mean_nomed	83
	mean	63
Mouse events	Mean_nomed	79
	mean	62
n-gram space	Mean_nomed	79
	mean	77

of an out-of-the-box classifier is intended to demonstrate that these results can be achieved in a straightforward manner.

Table 6.1 shows the accuracy of our completion time predictions, calculated via leave-one-out cross validation. Both state and edge space provide strong completion-time prediction results, with maximum accuracies of 83%. However, these classifiers can only take into account high-level changes in the software, as opposed to the lower-level physical actions that may characterize different participants, which leads us to investigate different encodings for further analyses.

6.3.2   EVENT-BASED ANALYSIS

Users move their mouse throughout the process of working with a visual analytic system. Sometimes they move the mouse purposefully, e.g., to click on a control, other times they hover over regions of interest, and sometimes they make idle movements. Where the state and edge space encodings fail to make use of this information, the event-based data encoding described in this section derives from the most raw interaction information available to capture innate behavioral differences.

Previous machine learning work has shown that mouse event data contains enough information to re-authenticate users for security purposes [109, 137, 138, 178]. We adapted the data representation of Pusara et al. [137] for our interaction data by calculating their set of statistics over event information. Because we are predicting completion time, we removed any statistics that we found to be correlated with completion time. Table 6.2 shows the set of functions we used to encapsulate the participants' mouse movements and raw clicks. This set includes statistics on click information (number of clicks and time between clicks), raw button click information

Table 6.2: Features calculated for SVM analysis of mouse movement and raw mouse click data. μ, σ, and μ'_3 refer to the mean, standard deviation, and third statistical moment. Pairwise indicates functions of pairs of consecutive events.

Click Event Features	Move Event Features
Clicks per second	Movement per second
Average time between clicks	Pairwise Euclidean distance (μ, σ, μ'_3)
% left, % right	Pairwise x distance (μ, σ, μ'_3)
% up, % down	Pairwise y distance (μ, σ, μ'_3)
% zoom in, % zoom out	Pairwise speed (μ, σ, μ'_3)
% found, % quit	Pairwise angle (μ, σ, μ'_3)
% clicks on image	

(percentage of clicks on a particular button, e.g., "% Left" refers to the percentage of button clicks on the "Pan Left" button), and derived mouse movement information (such as the number of moves, and the mean, standard deviation, and third statistical moment of the distance and angle between them). The set does not include total counts of clicks on different buttons or the total number of mouse movement events, because those were strongly correlated with the total completion time. In total, we use 27 features, listed across the 2 columns of Table 6.2.

As with the state-space representations, we apply SVMs to the mouse-event data. Table 6.1 shows the accuracy achieved with the mouse-event data using SVM classifiers, calculated using leave-one-out cross-validation. This approach manages a maximum score of 79%, which shows that there is strong signal in this low-level mouse data. The input features may reflect subconscious mouse movement habits more than actual intended actions, so the results indicate that the differences between populations may be driven by innate differences in approach or cognitive traits. Even though none of the features is individually correlated with the completion time, these low-level interaction statistics taken together are enough to analytically separate fast from slow users.

6.3.3 SEQUENCE-BASED ANALYSIS

The most direct representation of a user's process may be the sequence of direct interactions with software. Clicks are conscious actions that represent the user's intentions, and thus building classifiers based only on these semantically relevant interactions may provide more insight into why and how participants' analytical strategies differ. For our sequence-based analysis, we examine the sequences of button clicks used by participants to achieve the task of finding Waldo. We connect n-grams, a method from information retrieval for extracting short subsequences of words from collections of documents, to decision trees, a class of machine learning algorithms that produces human-readable classifiers.

N-Grams and Decision Trees

The n-gram method from information retrieval is intended for text, so an n-gram feature space must start with a string representation of data. We assign a unique symbol to each of the seven buttons in the interface: "L" for pan left, "R" for right, "U" for up, "D" for down, "I" for zoom in, "O" for out, and "F" for declaring Waldo found. Each participant's total interactions are thus given by an ordered string of symbols. We derive an n-gram vector space by considering each symbol a word, and each participant's sequence of words a document. Each dimension in the vector space then corresponds to one n-gram (i.e., one short sequence of user actions). Participants are represented by a vector of counts of the appearances of each n-gram in their interaction strings.

In our analyses we apply the J48 decision tree algorithm and NGramTokenizer from Weka [69] to classify participants based on task performance, and report accuracy scores from leave-one-out cross validation. The effectiveness of n-grams is sensitive to the choice of n. We empirically chose a combination of 2- and 3-grams as we found that to best balance accuracy and expressiveness of our eventual analytic output. Our results at classifying participants on completion time are shown in Table 6.1, revealing a top accuracy of 77% calculated with leave-one-out cross validation.

Decision Tree Interpretation

One advantage to using a decision tree with n-grams is that the resulting classifier is human-readable. Figure 6.4 shows the decision tree produced for the completion time data in n-gram space, using a mean split for classes. Each internal node shows a sequence of button clicks and the branches are labeled with the number of occurrences needed of that n-gram to take that branch. We can make several inferences about strategy from this tree. The root node indicates the strongest splitting criteria for the data. In this case, that node contains "L D," the n-gram corresponding to a participant clicking "Pan Left" then "Pan Down." If that sequence was clicked more than three times by anyone, that indicated the person would finish slowly. This makes sense because Waldo is in the upper right of the image. Moving in the wrong direction too many times can be expected to slow down progress at the task.

The "F U" and "D F R" nodes are also revealing. The "F" corresponds to telling the program that Waldo is found. These "F" button presses are not the last action, meaning they do not correspond to correctly finding Waldo. Instead, these sequences show participants' false guesses. Thus, the tree suggests that participants who made several false guesses finished the task more slowly.

Finally, the "O O I" and "L O I" nodes correspond to behavior where the participant zoomed out and then back in again. The "O I" component could indicate participants zooming

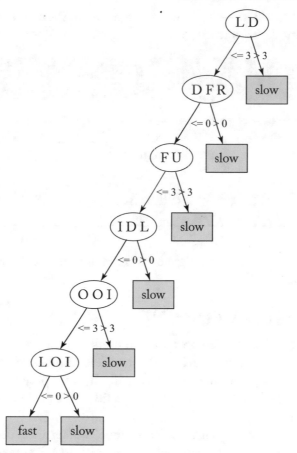

Figure 6.4: This is the decision tree generated as a classifier for fast vs. slow completion time with mean class splitting. Each internal node represents an individual decision to be made about a data point. The text within the node is the n-gram used to make the choice, and the labels on the out-edges indicate how to make the choice based on the count for a given data point of the n-gram specified. Leaf nodes indicate that a decision is made and are marked with the decided class.

out to gain context before zooming in again. Alternatively, the same subsequence could indicate participants zooming out and immediately back in, wasting time.

The readability of this technique shows promise for identifying trends in strategies and behaviors. We cannot guarantee that these interpretations reflect the participants' actual intentions, but rather submit these as possible reasons for what is shown in the tree. The real power of using n-grams and decision trees on interaction sequences is that it makes this type of hypoth-

Table 6.3: Personality Classifiers—all of these results are with SVM except when using n-grams, which we pair only with decision trees

Data Representation	Class Split	Accuracy (%)
LOC		
n-gram	Mean	67
Neuroticism		
Mouse events	Mean_nomed	62
Edge space	Mean_nomed	64
Extraversion		
Edge space	Mean	61

esizing possible, leading to deeper investigation when it is beneficial to understand how people are solving a given task.

6.4   PERSONALITY FINDINGS

The work in the previous chapters demonstrate that users will use a visualization system differently based on their personality traits. Motivated by these findings, we explore the efficacy of extracting personality traits from interactions. Specifically, we apply the same data encodings and machine learning algorithms used for the completion time analyses to predict users based on their personality traits.

Instead of classes derived from completion time, we separate users into *low* and *high* groups based on their scores on each personality inventory: LOC, extraversion, agreeableness, conscientiousness, neuroticism, and openness to experience. Consistent with our completion time analysis, we test both mean and mean_nomed splits (see Section 6.3). Table 6.3 summarizes our analysis results.

Across several techniques, we successfully classified users based on their LOC, neuroticism, and extraversion scores. Of the personality traits, our techniques were best with LOC, yielding classification accuracies as high as 67%. This supports the findings of Chapter 4 that of the personality traits, LOC was the strongest predictor of users' performance on visualization search tasks. Consistent with our findings, prior work also found significant effects with neuroticism and extraversion [66, 185].

6.5   LIMITED OBSERVATION TIME

The participants in our study were given as much time as they needed to complete the Waldo task. So far, the presented results have taken advantage of the full timespan of the collected data from their interactions to classify them. Investigating the minimal timespan required for

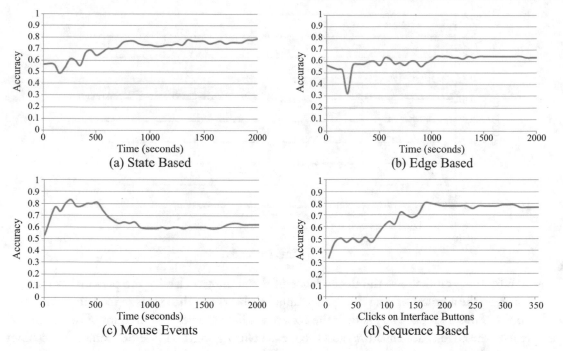

Figure 6.5: Graphs showing the ability to classify participants' completion time as a function of the extent of data collected. The x-axis represents the number of seconds of observation, or number of clicks for the sequence based data. The y-axis is the accuracy achieved after that amount of observation. Accuracy values are calculated with leave-one-out cross validation, and use the mean splitting method (see Section 6.3).

this type of analysis is crucial for potential real-time applications, so we evaluated our classifiers' performance as a function of the data collection time.

Figure 6.5 shows, for each of the different data representations, graphs of how task performance classification improves (on trend) with more observation time, i.e., more information available. Figure 6.6 shows one example of this behavior from personality trait classifiers. The x-axis is the amount of data collected, and the y-axis is the accuracy achieved by training and classifying with that amount of data. For all but the sequence-based analysis, the x-axis represents time. For the button click sequences, the x-axis is based on the number of clicks instead. Leave-one-out cross validation (LOOCV) and the mean-based class definition are used for all these results.

These graphs demonstrate two things. First, accuracy scores comparable to the final score can be achieved with much less than the maximum time. Note that close to the mean completion time, the encodings are achieving much of their eventual accuracy scores: state-based, 64% instead of its eventual 83%; edge-based, 60% compared to 63%; and sequence-based, 61%

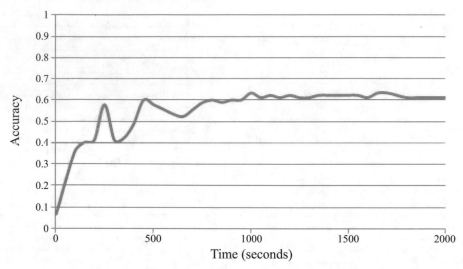

Figure 6.6: This graph shows the dependence of the ability to classify the personality trait extraversion on the amount of time the participants are observed. The x-axis represents the number of seconds of observation. The y-axis is the accuracy achieved after that amount of time. This example uses the edge space encoding and the mean splitting method (see Section 6.3). Accuracy values are calculated with leave-one-out cross validation.

as opposed to 77%. These correspond to 77%, 95%, and 79% of their final accuracy percentage scores, respectively.

Second, as expected, in most cases using more data allows for better results. In the case of the mouse event data, the accuracy peaks before reaching the average participant's finishing time, about 470 sec.

6.6   DISCUSSION AND FUTURE WORK

In this work, we have shown, via three encodings, that interaction data can be used to predict performance for real-time systems, and to infer personality traits. Our performance predictions ranged in accuracy from 62–83%. On personality traits, we were able to predict LOC, extraversion, and neuroticism with 61% up to 67% accuracy. Further, we found that with only two minutes of observation, i.e., a quarter of the average task completion time, we can correctly classify participants on performance at up to 95% of the final accuracy.

Given the above results, there are some fascinating implications and opportunities for future work. In this section, we discuss the choice of task and data representations, and how they may be generalized, differences between the personality results vs. those for completion time, and future work.

6.6.1   THE WALDO TASK AND OUR ENCODINGS

The Where's Waldo task was chosen because it is a generic visual search task. It is a simple example of an elementary sub-task that comes up often in visual analytics: looking for a needle in a haystack. The user can manipulate the view, in this case with simple controls, and employ strategy to meet a specific task objective. In this section we address how this experiment and our analyses may scale to other systems. Because our set of encodings is based on three core aspects of visual analytics, data, user, and interface, we frame the extensibility of our approach in terms of data and interface.

The data in our experiment is the image in which participants search for Waldo. At a data scale of twenty-megapixels, our state-based interaction encodings, which are closely tied to the data because they capture what parts of the image a participant sees, reach hundreds of features to over 1,000 features. As the size of the data (image) increases, the state space and edge space may not scale. However, the event-based and sequence-based encodings depend only on the interface, and thus could scale with larger image data.

Conversely, the interface in our experiment is a simple set of seven buttons. Increasing the complexity of the interface affects the event-based and sequence-based encodings. The mouse-event features include statistics about how often each button is pressed, and the sequence-based encoding requires a different symbol for each button. While these two encodings may not be able to scale to meet increased interface complexity, the state-based encoding is unaffected by the interface and thus could scale with the number of controls.

The three encodings we used in this chapter can all be extracted from the same interaction logs. Each one of them provides enough information to recover task performance efficiently. Because of their complementary relationships with the core concepts of interface and data, their strength, as an ensemble, at learning from interaction data is not strictly constrained by either interface or data complexity.

The scalability of the ensemble of encodings raises the possibility that our approach could be generalized to other visual search tasks and to complex visual analytics tasks. In particular, since users' interactions in visual analytics tasks have been shown to encode higher-level reasoning [47], we envision that our technique could be applied to other sub-tasks in visual analytics as well. Specifically, we consider the Waldo task as a form of the data search-and-filter task in the Pirolli and Card Sensemaking Loop [134]. We plan on extending our technique to analyzing user's interactions during other phases of the analytic process such as information foraging, evidence gathering, and hypothesis generation.

6.6.2   PERSONALITY

Being able to demonstrate that there is signal in this interaction data that encodes personality factors is exciting. However, none of the results for personality factors are as strong as those for completion time. Not only are the overall accuracy scores lower, but we found that in examining the time-based scores (as in Section 6.5), for many personality factors, there was not a persistent trend that more data helped the machine learning (Fig. 6.6 shows one of the stronger examples where there is a trend).

While the prediction accuracies are low, our results are consistent with prior findings [151] in the HCI community on individual differences research. Taken together, this suggests that although personality and cognitive traits can be recovered from users' interactions, the signals can be noisy and inconsistent. In order to better detect these signals, we plan to: (1) explore additional machine learning techniques, like boosting [144] for leveraging multiple learners together; and (2) apply our techniques to examine interactions from more complex visual analytics tasks. We expect the latter to amplify results as Chapter 4 have shown that personality trait effects are dampened when the task is simple. In Chapter 4, for complex inferential tasks, the effects were more pronounced and potentially easier to detect.

6.6.3   FUTURE WORK

This work is a first step in learning about users live from their interactions, and leaves many exciting questions to be answered with further research. The ability to classify users is interesting on its own, but an adaptive system could test the feasibility of applying this type of results in real time. For example, since locus of control can affect how people interact with visual representations of data (Chapter 4), a system that could detect this personality trait could adapt by offering the visualization expected to be most effective for the individual user. Different cognitive traits may prove more fruitful for adaptation, but even completion time could be used to adapt, by giving new users advice if they start to follow strategies that would lead to their classification as slow.

Further, of the data representations we evaluated, only the mouse events, the lowest-level interactions, encode any information about time passing during the task. The other representations do not encode the time between states or button presses, but that information could be useful for a future study. For our sequence-based analysis, our approach was to pair n-grams with decision trees for readability, but there are plenty of existing treatments of sequence data that remain to be tried for this type of data classification on visual analytic tasks, including sequence alignment algorithms, and random process models, e.g., Markov models. Finally, in this work we focused on encoding one aspect of data or interface at a time, but combining feature spaces could be powerful. In fact, in experimenting with a feature space that leverages multiple types of encodings, we achieved 96% accuracy on completion time with mean_nomed splitting.[1]

The experimental task was a simple version of a basic visual analytics sub-task. Our results could be strengthened by expanding the experiment to test Waldo in different locations, or different stimuli like maps with buildings and cars. The breadth of applicability could be evaluated by testing other elementary visual analytics tasks such as using tables to find data or comparing values through visual forms.

---

[1]Specifically, we tested a modified state space encoding where the zoom level information is replaced by an identifier of the button click that caused the state.

Our plans to extend this work expand on three fronts: (1) evaluating additional personal traits, like cognitive factors such as working memory, to our analyses; (2) trying further machine learning algorithms and encodings to learn from more of the information being collected, like the times of the interactions; and (3) extending experiments with different tasks including deeper exploration of visual search. We believe there are many opportunities to extend this work, both experimentally and analytically.

6.7   SUMMARY

In this chapter, we presented results of an online experiment we conducted where we recorded participants' mouse interactions as they played the game Where's Waldo. We broke the users into groups by how long it took them to find Waldo (completion time) and their personality traits. Visualizing the participants views of the data, we showed that there are differences in strategies across groups of users. We then applied machine learning techniques, and demonstrated that we can accurately classify the participants based on their completion time using multiple representations of their interactions: visualization states, low-level mouse events, and sequences of interface button clicks. By examining artifacts of our machine learning work with these sequences, we were able to identify short subsequences of interactions that identify groups of users. These human-readable classifier results hint at user strategies across groups. We were also able to detect and classify the participants based on some personality factors: LOC, extraversion, and neuroticism. Finally, we showed the dependence of the machine learning results on the observation time of the participants.

6.8   EXTENDED RESULTS

Although we demonstrated that completion time and personality can be modeled from raw interactions can be done directly with off-the-shelf tools using default settings, further attention to detail can yield stronger classifiers. In this appendix, we discuss some additional results that we achieved by tuning the algorithms, including applying principal component analysis (PCA) and optimizing the parameters of SVMs.

The SVM algorithm is sensitive to a slack parameter [38] and to the choice of kernel. Common practice is to address this by using a parameter search to find the best parameter values [74]. In the context of deploying the best possible classifier for a given dataset, that entails simply trying different choices of the parameter (or sets of parameters) and evaluating the classifiers until the best can be reported. Since our goal is instead to evaluate the classifiers and encodings themselves for this type of data, we take the approach of validating the algorithm of classifier+param-search. As usual for cross validation, the data is split into k folds. Each fold takes a turn as the test data, while the other folds are used for training, providing k samples of accuracy to be averaged for a total score. In testing a classifier+param-search algorithm, the algorithm being evaluated on one fold is one that chooses a parameter by testing which value

Table 6.4: Additional SVM result—all results are calculated using leave-one-out cross validation

Data Representation	Class Split	Classifier	Accuracy (%)
Completion Time			
Edge space	Mean_nomed	SVM_{poly}	87
	Mean	SVM_{poly}	72
Mouse events	Mean_nomed	SVM_{puly}	88
	Mean	SVM_{poly}	82
LOC			
Edge space	Mean	SVM_{poly}	62
State space	Mean_nomed	SVM_{poly}	63
State Space$_{PCA}$	Mean_nomed	SVM	63
Neuroticism			
Edge space	Mean_nomed	SVM_{poly}	68
State Space$_{PCA}$	Mean_nomed	SVM	68

produces the best classification result. To evaluate "best classification result," another (nested) cross validation is needed. The original fold's training data is split into folds again and cross validation is used over those inner folds to pick the optimal parameter. Weka implements a more sophisticated version of this practice that allows optimizing two parameters at once (generally referred to as grid search) and uses optimizing heuristics to limit the number of evaluations [131]. We have used this implementation to run a grid search that optimizes over (1) slack parameter and (2) degree of polynomial for kernel (1 or 2, i.e., linear or quadratic). In Table 6.4, this classifier is called SVM_{poly}. This table shows highlights of the results that we produced with this technique.

Another helpful factor in working with SVMs on high-dimensional data is principal component analysis. PCA projects the high-dimensional data into a lower-dimensional space defined by the eigenvectors of the original data. The number of eigenvectors is chosen to make sure that 95% of the variance in the data is accounted for in the low-dimensional approximation.[2] Applying PCA to the data space was particularly helpful in data representations like state space, which has a high degree of dimensionality. In Table 6.4, data representations with PCA applied are indicated by the subscript PCA.

Overall, the results in Table 6.4 show cases in which our tuning produced higher-accuracy classifiers, and revealed signal with feature spaces or class splitting criteria that otherwise could

---

[2]We used the Weka filter implementation with this option enabled.

not encode certain traits. The completion time results for the edge space and mouse event feature spaces are improvements of up to 32%. Specifically with edge space encoding and mean split, SVM_{poly} offers 82% accuracy instead of 62% with off-the-shelf SVM. In our earlier analyses, we did not find sufficient signal to report on LOC with any state-based encodings, but using PCA or parameter search makes that possible. Through applying standard methods for tuning SVM, we gained higher accuracy over our existing results, and demonstrated connections between encodings and traits that were otherwise obscured.

CHAPTER 7

The Adaptive User

> "To truly support the analytical reasoning process, we must enable the analyst to focus on what is truly important."
>
> James J. Thomas and Kristin A. Cook   [156]

Although we have laid a critical foundation for understanding the role of cognitive processes and individual differences in visualization, concretizing the intuition that each user experiences a visual interface through an individual cognitive lens is only half the battle. One promising next step is to calibrate the interface to each individual by intelligently adapting its design. Previous work has found that adaptive systems result in performance or satisfaction gains, when it responds to motor abilities [57], vision [57], or brain sensing [1, 150], among others. However, indiscriminately modifying the design of an interface may prevent the user from establishing a clear mental model of the system, decreasing the user's effectiveness and increasing feelings of loss of control.

To avoid this pitfall, it is important to have a thorough understanding of the connection between individual differences and performance on visualizations. In this chapter, we borrow techniques from Psychology and we investigate the impact of manipulating users' personality on observed behavior when using a visualization. We explore the idea of *nudging* a user [125] and demonstrate how priming techniques can be used to nudge a user's cognitive state and predictably influence performance. We also discuss design implications and the potential benefits of such a technique.

7.1   INTRODUCTION

Researchers in HCI and Psychology have investigated the efficacy of priming an individual's personality with the intent of temporarily influencing behavior. Studies include using emotionally-charged visual stimuli to inspire creativity [103] and eliciting varying levels of conformity by having users read words with positive or negative connotations [51]. Results indicate that non-invasive priming tasks can result in significant behavioral changes. If you wish to experience priming, take a look at Fig. 7.1. The image of a baby is an example of an affective priming stimulus, and studies suggest that exposure is likely to evoke a positive emotion.

Figure 7.1: **Photo by** Shelby Miller **on** Unsplash

In light of these findings, we posit that through priming, we may also be able to elicit performance changes when using visualization systems, specifically changes in speed and accuracy. We focus our study on the effects of priming a well-established personality trait known as *locus of control* (LOC). Research in personality psychology suggests that an individual's LOC may vary over time, and may even be intentionally manipulated [53, 85].

There exist many well-vetted techniques for priming personality factors. For example, Epley et al. used nonconscious priming to elicit conformity to social pressures [51]. Participants were given a scrambled sentence task containing words related to either conformity or rebellion, and were told that they were performing a pilot test. They found that this priming task was sufficient to elicit conformity in a subsequent social scenario. Chalfoun et al. introduced subliminal cues to enhance users' learning capabilities when using tutoring systems [25]. They found that by including positive cues such as hints to encourage inductive thinking, they were able to increase reasoning ability and improve decision making when solving logic problems.

Similar priming techniques have been used to manipulate the personality trait LOC. For example, Fisher and Johnstion [53] used priming to investigate the effects of psychological intervention targeting LOC in persons with disabilities. In this study, patients with chronic lower-back pain were randomly primed to score higher (measure more internal) or lower (measure more external) on the LOC scale. Researchers primed participants' LOC by asking experience recall questions: participants were asked to describe either times when they felt in control (thereby increasing their LOC score) or times they did not feel in control (thereby decreasing their LOC score). They found a significant difference in LOC scores before and after the application of this priming technique. They then assessed patients' perceived and actual physical ability using a lifting task. They reported that patients who were primed to be more internal spent more time on average performing the lifting task and selected heavier weights than the externally-primed group. Patients primed to be more external were significantly more likely to decline to participate in the lifting task. These results indicate that LOC can be primed using experience recall, and also demonstrates that priming LOC can influence behavior.

We hypothesize that we can significantly influence a user's speed and accuracy on visualization tasks by using priming techniques. Specifically, we expect that prompting an average user to be more internal will make them exhibit the behavior of internal participants from Chapter 4, and we expect a reverse effect when prompting an average user to be more external. Similarly, we posit that prompting internal participants to be more external and external participants to be more internal will lead to a reduction of differences between the groups, if not a full reversal of the original effect.

7.2   MANIPULATING LOCUS OF CONTROL

To test our hypotheses we replicated the study in Chapter 4, holding constant the views, datasets, and questions to enable us to make accurate comparisons between the two results. We conducted a targeted study extending the prior work by applying the experience recall techniques introduced by Fisher et al. [53] to manipulate participants' LOC. We measured participants' baseline LOC prior to the main task using a 23-point Rotter LOC Scale [141] and used this score to assign priming groups.

- Participants who scored **higher than 15** (designated *internal*) were given a task designed to **decrease** LOC score and assigned to group $I^{\rightarrow E}$. Participants of this group were expected to exhibit performance measures that are similar to the average participants reported in Chapter 4.

- Participants who scored **lower than 10** (designated *external*) were given a task designed to **increase** LOC score and assigned to the group $E^{\rightarrow I}$. These participants were also expected to exhibit performance measures that are similar to the average participants of the previous study.

- Participants who scored **between 10 and 15** (designated *average*) were randomly given a priming task and assigned to the appropriate group, either $A^{\rightarrow I}$ or $A^{\rightarrow E}$. We anticipated that average users who were primed internal or external would exhibit performance

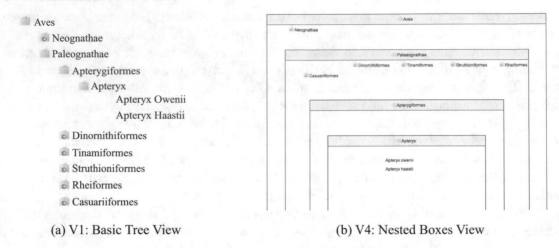

    (a) V1: Basic Tree View   (b) V4: Nested Boxes View

Figure 7.2: The two visualization used in the current study. They were two of the four visualizations designed in Chapter 4. V1 was designed to have a list-like metaphor while V4 was designed to be container-like.

measures similar to the internal users of the previous study, while those who were primed external would exhibit performance measures similar to the external users.

For simplicity, we used only the most extreme views (see Fig. 7.2) presented in Chapter 4, as their results were most compelling for these views. The order in which the views were presented was randomized, and participants were asked to complete a search and an inferential task for each view.

We recruited 300 participants via Amazon's Mechanical Turk service. Participants were assigned to one of two priming tasks which were slight modifications of those used by Fisher et al. [53]. In the first condition, participants were asked to describe times when they felt in control, which was designed to increase their LOC score (shifting their LOC toward *internal*). In the second condition, participants were asked to recall times when they felt they were not in control, which was designed to lower their LOC score (shifting their LOC toward *external*). Because our study was conducted on Mechanical Turk, the subjects completed priming tasks by entering three free-text examples of 100 words each to ensure effective priming. Below are the two priming stimuli used for this study.

Priming Question 1 (Increase Locus of Control)
"We know that one of the things that influence how well you can do everyday tasks is your sense of control over problems you face. The more control you believe you have, the better you will succeed at the things you try and do. If you feel optimistic and able to make the best of your situations, you will do very well. In the spaces provided below, give 3 examples of times when

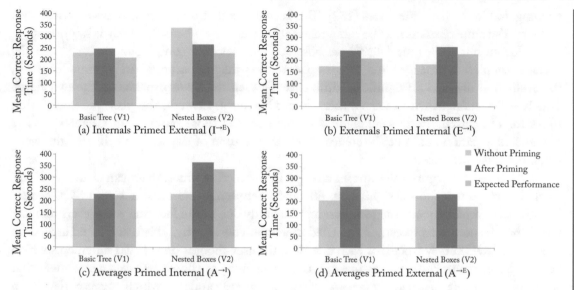

Figure 7.3: Mean correct response times on inferential task questions across the two views for each of the four priming groups. The average participants were successfully primed to behave as internal participants, while the internal and external participants were successfully primed to be more average.

you have felt in control and achieved things well. Each example must be at least 100 words long."

Priming Question 2 (Reduce Locus of Control)
"We know that one of the things that influence how well you can do everyday tasks is the number of obstacles you face on a daily basis. If you are having a particularly bad day today, you may not do as well as you might on a day when everything goes as planned. Variability is a normal part of life and you might think you can't do much about that aspect. In the spaces provided below, give 3 examples of times when you have felt out of control and unable to achieve something you set out to do. Each example must be at least 100 words long."

Our results confirm that we can successfully use priming to alter users' performance during complex tasks. We were able to both reduce differences between extreme user groups as well as create differences between average users using priming. It is important to note that the change in performance is both *statistically* and *practically* significant. By asking internal users to recall times when they did not feel in control we are able to effect a remarkable improvement in performance: the response gap between the two views for internal users were reduced from 110 seconds (roughly two minutes) to just 20 sec. Notably, we also found that priming affected a user's response time more than their accuracy. This is also true of previous work on metaphor priming and individual differences [182]. This is evidence that interactions between visual style and a user's frame of mind may be more relevant in situations where efficiency is important.

For all but one group ($A^{\rightarrow E}$) we observed the expected completion times. While the average completion time for the $A^{\rightarrow E}$ group was slower than the external users from Chapter 4, the evaluation revealed no significant difference between their performance on the two views. This is still consistent with external users from Chapter 4. Overall, the observed completion times for the current study were slightly slower than the those reported in Chapter 4. One possible explanation could be differences between the time of day and the time of the week when the tasks were posted.

Although we reported significant changes in completion times, the mean change in F was small for groups that showed significant differences between pre-test and post-test LOC scores. Feedback from participants suggests this may be partly due to the fact that some users remembered their responses from earlier and tried to answer consistently. This could be an artifact of the Rotter LOC survey where for each question, the user chooses one of two statements which best describes them. For instance, one question asked users to choose between a statement that suggests that *there will always be someone who doesn't like you* and one which suggests that *if you can't get people to like you then you don't understand how to get along with others*. While we believe priming does affect perceived control of one's environment, the Rotter LOC may not be the appropriate tool for measuring small changes. Future work could investigate the use of other LOC surveys which uses a Likert scale instead of a forced-choice scale.

To some extent, these results complicate previous findings on personality effects in visualization. We found no evidence that we are able to prime users with an extreme baseline LOC to adopt the behaviors of the other extreme using this technique. However, we were able to successfully prompt extreme users to exhibit behavior similar to averages. It is possible that users with a strong tendency in a certain personality trait can only be coaxed out of that tendency to a limited degree. This indicates that while volatile individual differences are still important factors to consider during design and evaluation, priming is not a panacea. In addition to furthering our understanding of the role of individual differences in future applications, these findings also shed light on the results of previous studies. In the next section, we discuss how this study relates to the design and evaluation of visualizations.

7.3   IMPLICATIONS

This and previous studies underscore that evaluating tools to help people think is a complicated endeavor. Our results suggest that traditional efficiency measures of speed and accuracy may not capture all of what we value in a visualization. While accuracy alone may not reflect the actual difficulty of a task, interaction time proves to be far too sensitive to minor changes in user inclination to provide generalizable information about a system. Evaluation must therefore go beyond simply analyzing the efficiency of a visual design but should also include methods that analyzes the user's cognitive factors.

7.3.1 EVALUATION

The way people think and solve problem is often situation-dependent. It is entirely possible that subtle aspects of user study procedures, task question design, and even a researcher's behavior can initiate unintentional cognitive priming and contribute a participant feeling more or less in control. If task performance can be affected by a user's cognitive state, this kind of unintentional priming could harm the validity of evaluation results. This recalls Ziemkiewicz and Kosara's finding [186] that metaphors used in the wording of task questions can interact with visualization layout in an evaluation setting.

Priming can also be intentional. While we only focused on LOC for this study, previous work highlights how other cognitive states can also affect performance on visualization systems [71, 185]. In some cases, the interaction of cognitive states can negatively affect both a user's speed and accuracy and therefore negatively affect evaluation. One practical application of priming is that we may be able to negate these disadvantages. Before evaluating visualizations, researchers can explicitly or subconsciously nudge the user into a specific frame of mind. By subtly presenting a positive news article as participants wait to begin the experiment or displaying a positive picture, researchers could affectively prime participants, making them better suited to perform certain tasks. Researchers can also administer an "unrelated" pre-task to disguise priming stimuli.

7.3.2 DESIGN

One possibility for visualization design is to prompt a user into a certain frame of mind better suited for the tool at hand. For example, verbal or textual elements such as instructions could also be tuned to temporarily prime the user, improving their capacity for working with a specific interface type. A system could use language in its instruction texts that primes the user to adopt a different frame. It may even be possible to design elements in a more subtle way. Lewis et al. demonstrated how the use of images can influence a user's emotional state [103]. Subtly including images in an interface design may also improve the performance of someone who is having a bad day. Indeed, such prompting may be implicitly at work in existing designs and may affect other cognitive states. Understanding this process better may make it possible to automate some of this design work.

Future systems can be equipped with a better understanding of users' cognitive states and automatically nudge users into a specific frame of mind. Priming may be used to counteract biases and encourage a user to experience an interface with a new perspective. They also proposed the use of priming in the design of collaborative systems. While is it often advantageous to have different perspectives, it is sometimes necessary for collaborators to see an interface as everyone else does. Priming has the potential to unify their conflicting perspectives when they exist.

That said, individual differences research remains necessary. One of our findings is that some users are simply more susceptible to prompting than others. Identifying these less flexible user groups and how they respond to varying visualization designs is still important if we are to completely understand how priming techniques can be used to control personality effects. While we have focused on LOC in this work, similar priming techniques exist for other cognitive states. These may also provide opportunities for controlling for individual differences within evaluation studies. However, it is first necessary to determine whether these cognitive states affect performance on visualization tasks.

7.4   SUMMARY

In this chapter, we demonstrated that by manipulating a user's LOC, we can prompt them to exhibit significantly different behavioral patterns. Our results also highlighted the sensitivity of evaluation measures such as response time to small situational changes. These findings help build toward understanding personality factors can affect the ways humans solve problems with visualizations and contribute to the development of systems that are robust to the effects of individual differences. This research also helps build toward a symbiosis between the system and the user, where not only do users adapt their systems to better suit their analytical needs, but systems can also encourage adaptation by the user to enhance performance.

CHAPTER 8

Future Directions and Open Challenges

"We have the opportunity to learn an enormous amount of new information from performing this work. And what is more, we will end up with a much stronger field that is based on solid foundations we can trust and build on."

Robert Kosara   [94]

This book contains much of the current research on individual differences in visualization use. We covered how individual differences can impact reasoning with real-world visualizations for medical risk communication, and saw that spatial ability influences reasoning across different representations of medical test statistics. We also saw that LOC is a dominant factor and is a significant predictor of speed, accuracy, and strategy with a variety of layout styles for hierarchical visualizations. In Chapters 5 and 6, we explored the feasibility of detecting user attributes from their interactions with a visualization tool. Altogether, the work in this book reveals how nuances in individual characteristics can determine the effectiveness of a visualization tool. We also saw that it is indeed possible to detect individual traits in real-time, thus supporting the promise of adaptive and personalized visualization systems.

The past few years have seen increasing demands for tools to explore and make sense of large or complex data. The popularity of ubiquitous and sensing devices, shared health decision-making, and the quantified-self movement, have led to an explosion of public data, exasperating the need for such visualization tools. Still, translation of research and expert systems to the general public has proven to be non-trivial. For example, studies show that people may not engage with expository visualizations as much as previously thought [18], forcing organizations such as the *New York Times* to reconsider the cost of building visualizations to supplement their text [159]. As visualization gains widespread importance, the tasks, scenarios, and people that visualizations must support are becoming more varied. Ultimately, making sense of visualization requires understanding how users vary and why.

8.1   NEXT STEPS

Visualization users differ greatly in experiences, backgrounds, personalities, and cognitive abilities, yet visualizations, like much other software, continue to be designed for a single ideal user. It would be clearly impractical to design each visualization for an individual user. However, knowledge of broad differences between user groups could be used to guide design for specific domains and to suggest multiple analysis modes or customization options in a single system. The body of work on individual differences in visualization provides a foundation for achieving this goal. However, successfully translating research to practice warrants more work.

It is currently difficult for ordinary developers, with no background in visualization or social science research, to identify potential issues with their design choices. Perhaps the best-supported cognitive trait is color vision deficiency. There exist several designer tools for testing or verifying the color inclusiveness of a design [29, 32] or for selecting palettes that are color-blind safe [33]. A key future direction is to enable practitioners, with no individual differences research background, to foresee the effect of their designs. More investigation is needed so that we can provide clear guidelines for research and practitioners, and success in the research agenda could transform how we evaluate and design visualizations for different user groups, tasks, and domains. There are many open questions and challenges.

8.1.1   AUTOMATICALLY INFERRING TRAITS FROM INTERACTION

The research projects in this book all used psychological surveys to estimate a person's cognitive traits. In real-world scenarios, however, it may be unrealistic to expect users to be subjected to a deluge of forms. Discovering new and unobtrusive methods to capture cognitive state, trait, and experience/bias will ultimately drive research in individual cognitive differences. For example, in Chapter 6, we saw how we might detect user attributes by analyzing their click steam data. In the broader visualization community, we have seen increased interest in developing algorithms to the model users' behavior and in investigating how we can used these techniques to improve visualization tools (examples include [41], [122], and [168]).

Although the research on user modeling and individual differences have largely been separate, analyzing their intersection could open the doors for many exciting future work. One of the most critical insight individual differences can manifest in a variety of ways. For instance, analyzing the portions of the *data* explored by the user can indicate a user's expertise and biases [168]. Chapter 6 showed that analyzing *actions* (e.g., pans and zooms) uncovered differences that were mediated by user's LOC scores and personality traits [21]. Other work demonstrates that tracking *visual attention* via eye-gaze can reveal differences in people with varying perceptual speed and visual working memory [151]. Therefore, to successfully infer individual traits, future work must consider a comprehensive set of encodings that include actions, data, and visual features.

The ability to automatically infer personality traits and individual characteristics will open many opportunities for tailoring visualization systems to better suit the user. However, bridging the gap between visualization and personality psychology can raise serious privacy concerns. It is important to be aware of the potential ethical challenges ahead, and take socially responsive steps to mitigate the effects.

8.1.2   A CLEARER MAPPING BETWEEN INDIVIDUAL DIFFERENCES AND TASK

Task design is critical to the success of an evaluation [114], and researchers have created taxonomies for the types of tasks and interactions that are feasible for a given visualization (for example [148], [177], and [180]). For future work, it is essential to recognize that "exploration" as a task carries several different meanings. Recent work by Battle and Heer [12] distinguishes between bottom-up exploration and top-down exploration. Bottom-up explorations "are driven in reaction to the data" [6] or "may be triggered by salient visual cues" [104]. This type of exploration is open-ended and the user's instincts largely drives the interactions. Top-down explorations, on the other hand, are based on a high-level goals or hypothesis [12, 65, 104].

One shortcoming of the prior work that investigates how individual traits impact exploration paths is they study only goal-driven search tasks. Because of this limitation, we know only the effect that individual traits have on interactions for top-down exploratory data analysis. We need systematic studies to investigate the correlation between task type and patterns of interactions, and how individual traits may mediate observations.

8.1.3   GENERALIZE ACROSS VISUALIZATION DESIGN

When we take a closer look at the previous results, much of the observed patterns of behavior can be explaining by local vs. global level precedence in processing information. For instance, when searching a tree visualization, participants with an external LOC (*Externals*) were more likely to perform a depth-first search while participants with an internal LOC (*Internals*) were more likely to perform a breadth-first search [128]. A depth-first search strategy suggests a local precedence information processing while a breadth-first search indicates a attention to global features and their relationships. As a result, our preliminary work demonstrated that Externals were faster and more accurate with indented tree visualization. It is possible that the design encourages a local exploration. Similarly, when searching for Waldo, we found that Externals were more likely explore at a lower zoom level, paying attention to local features, while Internals tended to only zoom when they believed they had identified the target. This preference for attending to local vs. global features suggests a pattern of behavior that may generalize across visualization designs. Future work is needed to investigate the relationship between individual traits and processing precedence across different designs.

8.1.4   REAL-TIME ADAPTATION

One important advantage of understanding individual users' cognitive states, traits, and biases as a cohesive structure is that this opens up the possibility of developing adaptive, mixed-initiative visualization systems [156]. Similar mixed-initiative systems were proposed in the HCI community by Horvitz in 1999 [73]. As noted by Thomas and Cook in *Illuminating the Path* [156], an important direction in advancing visual analytics research is the development of an automated, computational system that can assist a user in performing analytical tasks. However, most visualization systems today are designed in a one-size-fits-all fashion without the ability to adapt to different users' analytical needs into the design.

Creating such mixed-initiative visualization systems is particularly difficult as visualization are often designed to support complex thought and decision-making. Still, there is some evidence that successful adaptive systems can significantly improve a user's ability in performing complex tasks. In the recent work by Solovey et al. [150], the authors show that with the use of a brain imaging technology (fNIRS) to detect a user's cognitive states the system can adapt the amount of automation and notably improve the user's ability in performing a complicated robot navigation task. Afergan et al. also used fNIRS to detect mental states of unmanned aerial vehicles (UAVs) operators as the completed complex UAV navigation tasks [1]. They designed a system that successfully detects when the operator is in the state of boredom or high workload, and automatically increase or reduce workload depending on the operator's cognitive load [1].

It is clear that adaptive systems can offer new possibilities for visualization research and development [64], but more work is necessary to model *how* and *when* a system should adapt to a user's needs. In general, there are two ways to perform real-time system adaptations: (1) **dynamic back-end adaptations** and (2) **dynamic front-end adaptations**. Both concepts have existed for decades in the Artificial Intelligence and HCI communities [81].

Dynamic Back-End Adaptation

This is an unobtrusive approach to adaptation. In these systems, the current display remain unchanged but the system performs additional tasks in the background to support the user. For instance, a system could predict a user's search strategy and perform pre-fetching or pre-computation when the data are large. Existing work in the Database community shows that this is possible. Battle et al. demonstrated how pre-fetching mechanisms can be informed through analyzing users' interactions [11]. The system can also provide help or addition information in a separate window. From the HCI community, the Ambient Help [108] system displays information that may be relevant to a user's current task, and work by Billsus et al. [15] suggests improving such proactive information systems by allowing users to adjust their obtrusiveness.

Dynamic Front-End Adaptation

In these systems, the display will dynamically adjust itself to the support a specific user or task. For instance, to support exploratory tasks a system could assist a user by highlighting exploration paths that a user may be unlikely to explore, or recommend visualizations that are better suited for a detected analysis pattern. Relevant to this type of adaption, Sears and Schneiderman proposed split menus where the most frequently used menu items would percolate to the top of the list in order to facilitate faster access [145]. A popular example of similar menu adaptation is the Microsoft Smart Menus which was introduced in the Windows 2000 operating system. Also the in HCI community, Jefferson and Harvey successfully demonstrated how a system's graphical presentation can be adapted based on a user's color blindness [83, 84]. Recent work by Carenini et al. explored different ways of adapting information visualizations and demonstrated how interventions can significantly improve performance [23].

By monitoring the user and intelligently tailoring situationally appropriate information to the user, we can create next generation user interfaces that better support the user's analytics process. Furthermore, we can begin to design visualization systems that actually leverage the acuity of the human visual system as well as our capacity to understand and reason about complex data. Such systems may be able to overcome (or even leverage) some of the limitations imposed by the human brain such as limited working memory, bias, and fatigue.

Bibliography

[1] Daniel Afergan, Evan M. Peck, Erin T. Solovey, Andrew Jenkins, Samuel W. Hincks, Eli T. Brown, Remco Chang, and Robert J. K. Jacob. Dynamic difficulty using brain metrics of workload. In *Proc. ACM CHI*, ACM Press, Citeseer, 2014. DOI: 10.1145/2556288.2557230 67, 78

[2] Icek Ajzen. *Attitudes, Personality, and Behavior*. McGraw-Hill Education, UK, 2005. 9

[3] Bryce Allen. Individual differences and the conundrums of user-centered design: Two experiments. *Journal of the American Society for Information Science*, 51(6):508–520, 2000. DOI: 10.1002/(sici)1097-4571(2000)51:6<508::aid-asi3>3.0.co;2-q xii, 12

[4] Gordon Willard Allport. *Personality: A Psychological Interpretation*. 1937. 13

[5] Basak Alper, Nathalie Henry Riche, Fanny Chevalier, Jeremy Boy, and Metin Sezgin. Visualization literacy at elementary school. In *Proc. of the CHI Conference on Human Factors in Computing Systems*, pp. 5485–5497, ACM, 2017. DOI: 10.1145/3025453.3025877 4

[6] Sara Alspaugh, Nava Zokaei, Andrea Liu, Cindy Jin, and Marti A. Hearst. Futzing and moseying: Interviews with professional data analysts on exploration practices. *IEEE Transactions on Visualization and Computer Graphics*, 25(1):22–31, 2018. DOI: 10.1109/tvcg.2018.2865040 77

[7] Carl R. Anderson. Locus of control, coping behaviors, and performance in a stress setting: A longitudinal study. *Journal of Applied Psychology*, 62(4):446–451, 1977. DOI: 10.1037/0021-9010.62.4.446 13

[8] E. W. Anderson, K. C. Potter, L. E. Matzen, J. F. Shepherd, G. A. Preston, and C. T. Silva. A user study of visualization effectiveness using EEG and cognitive load. In *Computer Graphics Forum*, 30:791–800, Wiley, Online Library, 2011. DOI: 10.1111/j.1467-8659.2011.01928.x 10

[9] F. Gregory Ashby, Vivian V. Valentin, and A. U. Turken. The effects of positive affect and arousal on working memory and executive attention. *Advances in Consciousness Research*, 44:245–288, 2002. DOI: 10.1075/aicr.44.11ash 4

[10] Nuray M. Aykin and Turgut Aykin. Individual differences in human-computer interaction. *Computers and Industrial Engineering*, 20(3):373–379, 1991. DOI: 10.1016/0360-8352(91)90009-u 9

[11] Leilani Battle, Remco Chang, and Michael Stonebraker. Dynamic prefetching of data tiles for interactive visualization. *ACM Special Interest Group on Management of Data (SIGMOD)*. DOI: 10.1145/2882903.2882919 78

[12] Leilani Battle and Jeffrey Heer. Characterizing exploratory visual analysis: A literature review and evaluation of analytic provenance in tableau. In *Computer Graphics Forum*, 38:145–159, Wiley, Online Library, 2019. DOI: 10.1111/cgf.13678 77

[13] David Benyon and Dianne Murray. Developing adaptive systems to fit individual aptitudes. In *Proc. of the 1st International Conference on Intelligent user Interfaces*, pp. 115–121, ACM, 1993. DOI: 10.1145/169891.169925 9

[14] Jacques Bertin. *Semiology of Graphics: Diagrams, Networks, Maps.* 1983. xii

[15] Daniel Billsus, David M. Hilbert, and Dan Maynes-Aminzade. Improving proactive information systems. In *Proc. of the 10th International Conference on Intelligent user Interfaces*, pp. 159–166, ACM, 2005. DOI: 10.1145/1040830.1040869 78

[16] John Boddy, Annabel Carver, and Kevin Rowley. Effects of positive and negative verbal reinforcement on performance as a function of extraversion-introversion: Some tests of Gray's theory. *Personality and Individual Differences*, 7(1):81–88, 1986. DOI: 10.1016/0191-8869(86)90111-x 6

[17] Katy Börner, Adam Maltese, Russell Nelson Balliet, and Joe Heimlich. Investigating aspects of data visualization literacy using 20 information visualizations and 273 science museum visitors. *Information Visualization*, 15(3):198–213, 2016. DOI: 10.1177/1473871615594652 4

[18] Jeremy Boy, Francoise Detienne, and Jean-Daniel Fekete. Storytelling in information visualizations: Does it engage users to explore data? In *Proc. of the 33rd Annual ACM Conference on Human Factors in Computing Systems*, pp. 1449–1458, 2015. DOI: 10.1145/2702123.2702452 75

[19] Jeremy Boy, Ronald A. Rensink, Enrico Bertini, and Jean-Daniel Fekete. A principled way of assessing visualization literacy. *IEEE Transactions on Visualization and Computer Graphics*, 20(12):1963–1972, 2014. DOI: 10.1109/tvcg.2014.2346984 4

[20] Gary L. Brase. Pictorial representations in statistical reasoning. *Applied Cognitive Psychology*, 23(3):369–381, 2009. DOI: 10.1002/acp.1460 20

[21] Eli T. Brown, Alvitta Ottley, Helen Zhao, Quan Lin, Richard Souvenir, Alex Endert, and Remco Chang. Finding Waldo: Learning about users from their interactions. *IEEE Transactions on Visualization and Computer Graphics*, 20(1):2, 2014. DOI: 10.1109/tvcg.2014.2346575 xii, xvii, 10, 13, 15, 44, 76

[22] Tad T. Brunyé, Holly A. Taylor, David N. Rapp, and Alexander B. Spiro. Learning procedures: The role of working memory in multimedia learning experiences. *Applied Cognitive Psychology*, 20(7):917–940, 2006. DOI: 10.1002/acp.1236 25

[23] Giuseppe Carenini, Cristina Conati, Enamul Hoque, Ben Steichen, Dereck Toker, and James Enns. Highlighting interventions and user differences: Informing adaptive information visualization support. In *Proc. of the SIGCHI Conference on Human Factors in Computing Systems*, pp. 1835–1844, ACM, 2014. DOI: 10.1145/2556288.2557141 10, 79

[24] Simon Cassidy and Peter Eachus. Learning style, academic belief systems, self-report student proficiency and academic achievement in higher education. *Educational Psychology*, 20(3):307–320, 2000. DOI: 10.1080/713663740 13

[25] Pierre Chalfoun and Claude Frasson. Subliminal priming enhances learning in a distant virtual 3D intelligent tutoring system. *IEEE Technology and Engineering Education (ITEE)*, 3(4):125–130, 2008. 68

[26] Chaomei Chen. Individual differences in a spatial-semantic virtual environment. *Journal of the American Society for Information Science*, 51(6):529–542, 2000. DOI: 10.1002/(sici)1097-4571(2000)51:6<529::aid-asi5>3.0.co;2-f 10

[27] Chaomei Chen and Mary Czerwinski. Spatial ability and visual navigation: An empirical study. *New Review of Hypermedia and Multimedia*, 3(1):67–89, 1997. DOI: 10.1080/13614569708914684 xii, 4, 10, 12, 15, 20

[28] William S. Cleveland and Robert McGill. Graphical perception: Theory, experimentation, and application to the development of graphical methods. *Journal of the American Statistical Association*, 79(387):531–554, 1984. DOI: 10.1080/01621459.1984.10478080 xii

[29] Coblis: Color Blindness Simulator. https://www.color-blindness.com/coblis-color-blindness-simulator/. Accessed July 15, 2019. 76

[30] Cheryl A. Cohen and Mary Hegarty. Individual differences in use of external visualisations to perform an internal visualisation task. *Applied Cognitive Psychology*, 21(6):701–711, 2007. DOI: 10.1002/acp.1344 xii, 12, 15

[31] William G. Cole. Understanding Bayesian reasoning via graphical displays. In *ACM SIGCHI Bulletin*, 20:381–386, 1989. DOI: 10.1145/67450.67522 20

[32] Color Oracle. `https://colororacle.org`. Accessed July 15, 2019. 76

[33] ColorBrewer 2.0. `http://colorbrewer2.org/`. Accessed July 15, 2019. 76

[34] Cristina Conati, Giuseppe Carenini, Enamul Hoque, Ben Steichen, and Dereck Toker. Evaluating the impact of user characteristics and different layouts on an interactive visualization for decision making. In *Computer Graphics Forum*, 33:371–380, Wiley, Online Library, 2014. DOI: 10.1111/cgf.12393 xii, 10, 12, 15

[35] Cristina Conati, Giuseppe Carenini, Dereck Toker, and Sébastien Lallé. Towards user-adaptive information visualization. In *AAAI*, pp. 4100–4106, 2015.

[36] Cristina Conati and Heather Maclaren. Exploring the role of individual differences in information visualization. In *Proc. of the Working Conference on Advanced Visual Interfaces*, pp. 199–206, ACM, 2008. DOI: 10.1145/1385569.1385602 xi, xii, 4, 10, 12, 15

[37] Kristin A. Cook and James J. Thomas. Illuminating the path: The research and development agenda for visual analytics. *Technical Report*, Pacific Northwest National Lab. (PNNL), Richland, WA, 2005. xi

[38] Corinna Cortes and Vladimir Vapnik. Support-vector networks. *Machine Learning*, 20(3):273–297, 1995. DOI: 10.1007/bf00994018 63

[39] Leda Cosmides and John Tooby. Are humans good intuitive statisticians after all? Rethinking some conclusions from the literature on judgment under uncertainty. *Cognition*, 58(1):1–73, 1996. DOI: 10.1016/0010-0277(95)00664-8 21

[40] R. Jordan Crouser, Alvitta Ottley, and Remco Chang. Balancing human and machine contributions in human computation systems. In *Handbook of Human Computation*, pp. 615–623, Springer, 2013. DOI: 10.1007/978-1-4614-8806-4_48 47

[41] Filip Dabek and Jesus J. Caban. A grammar-based approach for modeling user interactions and generating suggestions during the data exploration process. *IEEE Transactions on Visualization and Computer Graphics*, 23(1):41–50, 2017. DOI: 10.1109/tvcg.2016.2598471 76

[42] Rossana De Beni, Francesca Pazzaglia, Valerie Gyselinck, and Chiara Meneghetti. Visuospatial working memory and mental representation of spatial descriptions. *European Journal of Cognitive Psychology*, 17(1):77–95, 2005. DOI: 10.1080/09541440340000529 12

[43] Andrew Dillon and Charles Watson. User analysis in HCI: The historical lesson from individual differences research. *International Journal of Human-Computer Studies*, 45:619–637, 1996. DOI: 10.1006/ijhc.1996.0071 9

[44] Evanthia Dimara, Anastasia Bezerianos, and Pierre Dragicevic. The attraction effect in information visualization. *IEEE Transactions on Visualization and Computer Graphics*, 23(1):471–480, 2016. DOI: 10.1109/tvcg.2016.2598594 4

[45] Evanthia Dimara, Steven Franconeri, Catherine Plaisant, Anastasia Bezerianos, and Pierre Dragicevic. A task-based taxonomy of cognitive biases for information visualization. *IEEE Transactions on Visualization and Computer Graphics*, 2018. DOI: 10.1109/tvcg.2018.2872577 4

[46] Courtney C. Dornburg, Laura E. Matzen, Travis L. Bauer, and Laura A. McNamara. Working memory load as a novel tool for evaluating visual analytics. In *IEEE Symposium on Visual Analytics Science and Technology*, pp. 217–218, 2009. DOI: 10.1109/vast.2009.5333468 10

[47] Wenwen Dou, Dong Hyun Jeong, Felesia Stukes, William Ribarsky, Heather Richter Lipford, and Remco Chang. Recovering reasoning process from user interactions. *IEEE Computer Graphics and Applications*, 2009. DOI: 10.1109/mcg.2009.49 xii, 48, 61

[48] David M. Eddy. Probabilistic reasoning in clinical medicine: Problems and opportunities. In D. Kahneman, P. Slovic, and A. Tversky, Eds., *Judgment Under Uncertainty: Heuristics and Biases*, pp. 249–267, 1982. DOI: 10.1017/cbo9780511809477.019 19

[49] Geoffrey Ellis. *Cognitive Biases in Visualizations*. Springer, 2018. DOI: 10.1007/978-3-319-95831-6 4

[50] Niklas Elmqvist, Andrew Vande Moere, Hans-Christian Jetter, Daniel Cernea, Harald Reiterer, and T. J. Jankun-Kelly. Fluid interaction for information visualization. *Information Visualization*, 10(4):327–340, 2011. DOI: 10.1177/1473871611413180 39

[51] Nicholas Epley and Thomas Gilovich. Just going along: Nonconscious priming and conformity to social pressure. *Journal of Experimental Social Psychology*, 35(6):578–589, 1999. DOI: 10.1006/jesp.1999.1390 67, 68

[52] Maureen J. Findley and Harris M. Cooper. Locus of control and academic achievement: A literature review. *Journal of Personality and Social Psychology*, 44(2):419–427, 1983. DOI: 10.1037/0022-3514.44.2.419 13

[53] Keren Fisher and Marie Johnstion. Experimental manipulation of perceived control and its effect on disability. *Psychology and Health*, 11(5):657–669, 1996. DOI: 10.1080/08870449608404995 14, 68, 69, 70

[54] Barbara L. Fredrickson. What good are positive emotions? *Review of General Psychology*, 2(3):300, 1998. DOI: 10.1037/1089-2680.2.3.300 4

[55] Hendrik Friederichs, Sandra Ligges, and Anne Weissenstein. Using tree diagrams without numerical values in addition to relative numbers improves students' numeracy skills a randomized study in medical education. *Medical Decision Making*, 2013. DOI: 10.1177/0272989x13504499 20

[56] Maria-Elena Froese, Melanie Tory, Guy-Warwick Evans, and Kedar Shrikhande. Evaluation of static and dynamic visualization training approaches for users with different spatial abilities. *IEEE Transactions on Visualization and Computer Graphics*, 19(12):2810–2817, 2013. DOI: 10.1109/tvcg.2013.156 10

[57] Krzysztof Gajos and Daniel S. Weld. Supple: Automatically generating user interfaces. In *Proc. of the 9th International Conference on Intelligent User Interfaces*, pp. 93–100, ACM, 2004. DOI: 10.1145/964442.964461 67

[58] Krzysztof Z. Gajos and Krysta Chauncey. The influence of personality traits and cognitive load on the use of adaptive user interfaces. In *Proc. of the 22nd International Conference on Intelligent User Interfaces*, pp. 301–306, ACM, 2017. DOI: 10.1145/3025171.3025192 9

[59] Rocio Garcia-Retamero and Mirta Galesic. Who profits from visual aids: Overcoming challenges in people's understanding of risks. *Social Science and Medicine*, 70(7):1019–1025, 2010. DOI: 10.1016/j.socscimed.2009.11.031 19

[60] Rocio Garcia-Retamero and Ulrich Hoffrage. Visual representation of statistical information improves diagnostic inferences in doctors and their patients. *Social Science and Medicine*, 83:27–33, 2013. DOI: 10.1016/j.socscimed.2013.01.034 20

[61] Gerd Gigerenzer, Wolfgang Gaissmaier, Elke Kurz-Milcke, Lisa M. Schwartz, and Steven Woloshin. Helping doctors and patients make sense of health statistics. *Psychological Science in the Public Interest*, 8(2):53–96, 2007. DOI: 10.1111/j.1539-6053.2008.00033.x 20

[62] Gerd Gigerenzer and Ulrich Hoffrage. How to improve bayesian reasoning without instruction: Frequency formats. *Psychological Review*, 102(4):684, 1995. DOI: 10.1037/0033-295x.102.4.684 20, 21

[63] Lewis R. Goldberg, John A. Johnson, Herbert W. Eber, Robert Hogan, Michael C. Ashton, C. Robert Cloninger, and Harrison G. Gough. The international personality item pool and the future of public-domain personality measures. *Journal of Research in Personality*, 40(1):84–96, 2006. DOI: 10.1016/j.jrp.2005.08.007 40

[64] David Gotz and Zhen Wen. Behavior-driven visualization recommendation. In *Proc. of the 14th International Conference on Intelligent User Interfaces*, pp. 315–324, ACM, 2009. DOI: 10.1145/1502650.1502695 78

[65] David Gotz and Michelle X. Zhou. Characterizing users' visual analytic activity for insight provenance. *Information Visualization*, 8(1):42–55, 2009. DOI: 10.1109/vast.2008.4677365 77

[66] Tera Marie Green and Brian Fisher. Towards the personal equation of interaction: The impact of personality factors on visual analytics interface interaction. In *IEEE Symposium on Visual Analytics Science and Technology (VAST)*, pp. 203–210, 2010. DOI: 10.1109/vast.2010.5653587 xii, 4, 10, 13, 14, 15, 29, 30, 32, 34, 39, 40, 52, 58

[67] Tera Marie Green, Dong Hyun Jeong, and Brian Fisher. Using personality factors to predict interface learning performance. In *System Sciences (HICSS), 43rd Hawaii International Conference on*, pp. 1–10, IEEE, 2010. DOI: 10.1109/hicss.2010.431 xii, 13, 14, 15, 29

[68] Anzu Hakone, Lane Harrison, Alvitta Ottley, Nathan Winters, Caitlin Gutheil, Paul K. J. Han, and Remco Chang. Proact: Iterative design of a patient-centered visualization for effective prostate cancer health risk communication. *IEEE Transactions on Visualization and Computer Graphics*, 23(1):601–610, 2016. DOI: 10.1109/tvcg.2016.2598588 20

[69] Mark Hall, Eibe Frank, Geoffrey Holmes, Bernhard Pfahringer, Peter Reutemann, and Ian H. Witten. The WEKA data mining software: An update. *ACM SIGKDD Explorations Newsletter*, 11(1):10–18, 2009. DOI: 10.1145/1656274.1656278 53, 56

[70] Martin Handford. *Where's Waldo?* Little, Brown, Boston, 1987. 48

[71] Lane Harrison, Drew Skau, Steven Franconeri, Aidong Lu, and Remco Chang. Influencing visual judgment through affective priming. In *Proc. of the SIGCHI Conference on Human Factors in Computing Systems*, pp. 2949–2958, ACM, 2013. DOI: 10.1145/2470654.2481410 73

[72] W. Trey Hill and Gary L. Brase. When and for whom do frequencies facilitate performance? On the role of numerical literacy. *The Quarterly Journal of Experimental Psychology*, 65(12):2343–2368, 2012. DOI: 10.1080/17470218.2012.687004 20

[73] Eric Horvitz. Principles of mixed-initiative user interfaces. In *Proc. of the SIGCHI Conference on Human Factors in Computing Systems*, pp. 159–166, ACM, 1999. DOI: 10.1145/302979.303030 78

[74] Chih-Wei Hsu, Chih-Chung Chang, and Chih-Jen Lin. A practical guide to support vector classification. *Technical Report*, National Taiwan University, Taipei 106, Taiwan, April 2010. 52, 63

[75] Weidong Huang, Peter Eades, and Seok-Hee Hong. Measuring effectiveness of graph visualizations: A cognitive load perspective. *Information Visualization*, 8(3):139–152, January 2009. DOI: 10.1057/ivs.2009.10 4

[76] Clark L. Hull. *Aptitude Testing*, 1928. 9

[77] Jessica Hullman and Nicholas Diakopoulos. Visualization rhetoric: Framing effects in narrative visualization. *Visualization and Computer Graphics, IEEE Transactions on*, 17(12):2231–2240, 2011. DOI: 10.1109/tvcg.2011.255 27

[78] Jessica Hullman, Steven Drucker, Nathalie Henry Riche, Bongshin Lee, Danyel Fisher, and Eytan Adar. A deeper understanding of sequence in narrative visualization. *Visualization and Computer Graphics, IEEE Transactions on*, 19(12):2406–2415, 2013. DOI: 10.1109/tvcg.2013.119 27

[79] Michel C. Ioannou, Karin Mogg, and Brendan P. Bradley. Vigilance for threat: Effects of anxiety and defensiveness. *Personality and Individual Differences*, 36(8):1879–1891, 2004. DOI: 10.1016/j.paid.2003.08.018 35

[80] Dan Ispas and Walter C. Borman. *Personnel Selection, Psychology of*. 2015. DOI: 10.1016/b978-0-08-097086-8.22014-x 10

[81] Julie A. Jacko. *Human Computer Interaction Handbook: Fundamentals, Evolving Technologies, and Emerging Applications*. CRC Press, 2012. DOI: 10.1201/b11963 78

[82] Robert J. K. Jacob, John J. Leggett, Brad A. Myers, and Randy Pausch. Interaction styles and input/output devices. *Behaviour and Information Technology*, 12(2):69–79, 1993. DOI: 10.1080/01449299308924369 47

[83] Luke Jefferson and Richard Harvey. Accommodating color blind computer users. In *Proc. of the 8th International ACM SIGACCESS Conference on Computers and Accessibility*, pp. 40–47, 2006. DOI: 10.1145/1168987.1168996 79

[84] Luke Jefferson and Richard Harvey. An interface to support color blind computer users. In *Proc. of the SIGCHI Conference on Human Factors in Computing Systems*, pp. 1535–1538, ACM, 2007. DOI: 10.1145/1240624.1240855 79

[85] Marie Johnston, Penny Gilbert, Cecily Partridge, and Jenny Collins. Changing perceived control in patients with physical disabilities: An intervention study with patients receiving rehabilitation. *British Journal of Clinical Psychology*, 31(1):89–94, 1992. DOI: 10.1111/j.2044-8260.1992.tb00972.x 68

[86] Timothy A. Judge and Joyce E. Bono. Relationship of core self-evaluations traits—self-esteem, generalized self-efficacy, locus of control, and emotional stability—with job satisfaction and job performance: A meta-analysis. *Journal of Applied Psychology*, 86(1):80–92, 2001. DOI: 10.1037/0021-9010.86.1.80 13, 36

[87] Daniel Keim, Gennady Andrienko, Jean-Daniel Fekete, Carsten Görg, Jörn Kohlham-mer, and Guy Melançon. Visual analytics: Definition, process, and challenges. In *Proc. of the IEEE Conference on Information Visualization*, pp. 154–175, Springer Berlin Hei-delberg, 2008. DOI: 10.1007/978-3-540-70956-5_7 47

[88] Vince J. Kellen. The effects of diagrams and relational complexity on user performance in conditional probability problems in a non-learning context. Doctoral Thesis, DePaul University, 2012. 20, 21

[89] Vince J. Kellen, Susy Chan, and Xiaowen Fang. Facilitating conditional probability prob-lems with visuals. In *Human-Computer Interaction. Interaction Platforms and Techniques*, pp. 63–71, Springer, 2007. DOI: 10.1007/978-3-540-73107-8_8 26

[90] Doreen Kimura.   *Sex and Cognition*.   MIT Press, 2000. DOI: 10.7551/mit-press/6194.001.0001 xii, 10

[91] Gary Klein, Jennifer K. Phillips, Erica L. Rall, and Deborah A. Peluso. A data-frame theory of sensemaking. *Expertise out of Context*, pp. 113–155, 2007. 39

[92] Kitty Klein and Adriel Boals. Expressive writing can increase working memory capac-ity. *Journal of Experimental Psychology: General*, 130(3):520, 2001. DOI: 10.1037/0096-3445.130.3.520 4

[93] Alfred Kobsa. User modeling: Recent work, prospects and hazards. *Human Factors in Information Technology*, 10:111–111, 1993. 10

[94] Robert Kosara. An empire built on sand: Reexamining what we think we know about visualization. In *Proc. of the 6th Workshop on Beyond Time and Errors on Novel Evaluation Methods for Visualization*, pp. 162–168, ACM, 2016. DOI: 10.1145/2993901.2993909 75

[95] Robert Kosara and Jock Mackinlay. Storytelling: The next step for visualization. *Com-puter*, 46(5):44–50, 2013. DOI: 10.1109/mc.2013.36 27

[96] B. C. Kwon, B. Fisher, and J. S. Yi.   Visual analytic roadblocks for novice in-vestigators.   *IEEE Visual Analytics Science and Technology*, pp. 3–11, 2011. DOI: 10.1109/vast.2011.6102435 3

[97] Sébastien Lallé, Cristina Conati, and Giuseppe Carenini. Impact of individual differences on user experience with a real-world visualization inteface for public engagement. In *Proc. of the 25th Conference on User Modeling, Adaptation and Personalization*, pp. 369–370, 2017.

[98] Sébastien Lallé, Dereck Toker, and Cristina Conati. Gaze-driven adaptive interventions for magazine-style narrative visualizations. *ArXiv Preprint ArXiv:1909.01379*, 2019. DOI: 10.1109/tvcg.2019.2958540 10

[99] Terran Lane and Carla E. Brodley. Temporal sequence learning and data reduction for anomaly detection. *ACM Transactions on Information and System Security*, 2(3):295–331, August 1999. DOI: 10.1145/288090.288122 48

[100] Bongshin Lee, Petra Isenberg, Nathalie Henry Riche, and Sheelagh Carpendale. Beyond mouse and keyboard: Expanding design considerations for information visualization interactions. *Visualization and Computer Graphics, IEEE Transactions on*, 18(12):2689–2698, 2012. DOI: 10.1109/tvcg.2012.204 47

[101] Sukwon Lee, Sung-Hee Kim, Ya-Hsin Hung, Heidi Lam, Youn-ah Kang, and Ji Soo Yi. How do people make sense of unfamiliar visualizations?: A grounded model of novice's information visualization sensemaking. *Visualization and Computer Graphics, IEEE Transactions on*, 22(1):499–508, 2016. DOI: 10.1109/tvcg.2015.2467195 xii

[102] Sukwon Lee, Sung-Hee Kim, and Bum Chul Kwon. Vlat: Development of a visualization literacy assessment test. *IEEE Transactions on Visualization and Computer Graphics*, 23(1):551–560, 2016. DOI: 10.1109/tvcg.2016.2598920 4

[103] Sheena Lewis, Mira Dontcheva, and Elizabeth Gerber. Affective computational priming and creativity. In *Proc. of the SIGCHI Conference on Human Factors in Computing Systems*, pp. 735–744, ACM, 2011. DOI: 10.1145/1978942.1979048 67, 73

[104] Zhicheng Liu and Jeffrey Heer. The effects of interactive latency on exploratory visual analysis. *IEEE Transactions on Visualization and Computer Graphics*, 20(12):2122–2131, 2014. DOI: 10.1109/tvcg.2014.2346452 77

[105] Zhicheng Liu, Nancy Nersessian, and John Stasko. Distributed cognition as a theoretical framework for information visualization. *IEEE Transactions on Visualization and Computer Graphics*, 14(6), 2008. DOI: 10.1109/tvcg.2008.121 xi

[106] Jock Mackinlay. Automating the design of graphical presentations of relational information. *ACM Transactions on Graphics (TOG)*, 5(2):110–141, 1986. DOI: 10.1145/22949.22950 xii

[107] Laura Martignon and Christoph Wassner. Teaching decision making and statistical thinking with natural frequencies. In *Proc. of the 6th International Conference on Teaching of Statistics. Ciudad del Cabo: IASE. CD ROM*, 2002. 20

[108] Justin Matejka, Tovi Grossman, and George Fitzmaurice. Ambient help. In *Proc. of the SIGCHI Conference on Human Factors in Computing Systems*, pp. 2751–2760, ACM, 2011. DOI: 10.1145/1978942.1979349 78

[109] Roy A. Maxion. Masquerade detection using enriched command lines. In *Proc. of the IEEE International Conference on Dependable Systems and Networks*, pp. 5–14, 2003. DOI: 10.1109/dsn.2003.1209911 48, 54

[110] Mary H. McCaulley. The Myers-Briggs type indicator: A Jungian model for problem solving. *New Directions for Teaching and Learning*, 30:37–53, 1987. DOI: 10.1002/tl.37219873005 36

[111] Luana Micallef, Pierre Dragicevic, and J. Fekete. Assessing the effect of visualizations on Bayesian reasoning through crowdsourcing. *Visualization and Computer Graphics, IEEE Transactions on*, 18(12):2536–2545, 2012. DOI: 10.1109/tvcg.2012.199 19, 20, 23

[112] Anthony B. Miller, Cornelia J. Baines, Ping Sun, Teresa To, and Steven A. Narod. Twenty five year follow-up for breast cancer incidence and mortality of the Canadian national breast screening study: Randomised screening trial. *BMJ: British Medical Journal*, 2014. DOI: 10.1097/ogx.0000000000000066 20

[113] Thomas M. Mitchell. *Machine Learning*, 1st ed., McGraw-Hill, Inc., New York, 1997. 52

[114] T. Munzner. A nested process model for visualization design and validation. *IEEE Visualization and Computer Graphics*, 15(6):921–928, 2009. DOI: 10.1109/tvcg.2009.111 xii, 77

[115] National Center for Biotechnology Information. NCBI genome database. http://www.ncbi.nlm.nih.gov/genome/. Accessed 11/18/2010. 32, 41

[116] Donald A. Norman and Stephen W. Draper. *User Centered System Design: New Perspectives on Human-Computer Interaction*. CRC Press, 1986. DOI: 10.1201/b15703 43

[117] Dennis W. Organ. Extraversion, locus of control, and individual differences in conditionability in organizations. *Journal of Applied Psychology*, 60(3):401, 1975. DOI: 10.1037/h0076627 6

[118] Rita Orji, Lennart E. Nacke, and Chrysanne Di Marco. Towards personality-driven persuasive health games and gamified systems. In *Proc. of the CHI Conference on Human Factors in Computing Systems*, pp. 1015–1027, 2017. DOI: 10.1145/3025453.3025577 9

[119] Alvitta Ottley. Toward personalized visualizations. Ph.D. thesis, Tufts University, 2016. xii

[120] Alvitta Ottley, R. Jordan Crouser, Caroline Ziemkiewicz, and Remco Chang. Priming locus of control to affect performance. In *IEEE Conference on Visual Analytics Science and Technology (VAST)*, pp. 237–238, 2012. DOI: 10.1109/vast.2012.6400535 xvii, 13

[121] Alvitta Ottley, R. Jordan Crouser, Caroline Ziemkiewicz, and Remco Chang. Manipulating and controlling for personality effects on visualization tasks. *Information Visualization*, 2013. DOI: 10.1177/1473871613513227 xii, 40, 52

[122] Alvitta Ottley, Roman Garnett, and Ran Wan. Follow the clicks: Learning and anticipating mouse interactions during exploratory data analysis. In *Computer Graphics Forum*. Wiley, Online Library, 2019. DOI: 10.1111/cgf.13670 76

[123] Alvitta Ottley, Aleksandra Kaszowska, R. Jordan Crouser, and Evan M. Peck. The curious case of combining text and visualization. *Short Paper Proceedings of the Eurographics/IEEE VGTC Symposium on Visualization (EuroVis)*, 2020. 20

[124] Alvitta Ottley, Blossom Metevier, Paul K. J. Han, and Remco Chang. Visually communicating Bayesian statistics to laypersons. In *Technical Report*. Tufts University, 2012. 20, 23

[125] Alvitta Ottley, Evan M. Peck, Lane Harrison, and Remco Chang. The adaptive user: Priming to improve interaction. *Workshop on proceeding Many People, Many Eyes: Aggregating Influences of Visual Perception on User Interface Design, Conference on Human Factors in Computing Systems (CHI)*, 2013. 67

[126] Alvitta Ottley, Evan M. Peck, Lane T. Harrison, Daniel Afergan, Caroline Ziemkiewicz, Holly A. Taylor, Paul K. J. Han, and Remco Chang. Improving Bayesian reasoning: The effects of phrasing, visualization, and spatial ability. *IEEE Transactions on Visualization and Computer Graphics*, 22(1):529–538, 2015. DOI: 10.1109/tvcg.2015.2467758 xvii, 10

[127] Alvitta Ottley, Evan M. Peck, Lane T. Harrison, Daniel Afergan, Caroline Ziemkiewicz, Holly A. Taylor, Paul K. J. Han, and Remco Chang. Improving Bayesian reasoning: The effects of phrasing, visualization, and spatial ability. *Visualization and Computer Graphics, IEEE Transactions on*, 22(1):529–538, 2016. DOI: 10.1109/tvcg.2015.2467758 xii

[128] Alvitta Ottley, Huahai Yang, and Remco Chang. Personality as a predictor of user strategy: How locus of control affects search strategies on tree visualizations. In *Proc. of the SIGCHI Conference on Human Factors in Computing Systems*. ACM, 2015. DOI: 10.1145/2702123.2702590 xii, xvii, 10, 13, 77

[129] Fred G. W. C. Paas and Jeroen J. G. Van Merriënboer. The efficiency of instructional conditions: An approach to combine mental effort and performance measures. *Human Factors*, 35(4):737–743, 1993. DOI: 10.1177/001872089303500412 4

[130] Evan M. Peck, Alvitta Ottley, Beste F. Yuksel, Remco Chang, and Lane Harrison. ICD 3: Towards a 3-dimensional model of individual cognitive differences. *ACM BELIV'12: Beyond Time and Errors: Novel Evaluation Methods for Information Visualization*, 2012. DOI: 10.1145/2442576.2442582 xvii, 4, 44

[131] Bernhard Pfahringer, Geoff Holmes, and Fracpete. Class gridsearch, revision 9733. http://weka.sourceforge.net/doc.stable/weka/classifiers/meta/GridSearch.html, March 2014. 64

[132] Elizabeth A. Phelps, Sam Ling, and Marisa Carrasco. Emotion facilitates perception and potentiates the perceptual benefits of attention. *Psychological Science*, 17(4):292–299, 2006. DOI: 10.1111/j.1467-9280.2006.01701.x 4

[133] William A. Pike, John Stasko, Remco Chang, and Theresa A. O'Connell. The science of interaction. *Information Visualization*, 8(4):263–274, 2009. DOI: 10.1057/ivs.2009.22 39, 47

[134] Peter Pirolli and Stuart Card. The sensemaking process and leverage points for analyst technology as identified through cognitive task analysis. In *Proc. of the International Conference on Intelligence Analysis*, 5:2–4, 2005. 39, 61

[135] Kym E. Pocius. Personality factors in human-computer interaction: A review of the literature. *Computers in Human Behavior*, 7(3):103–135, 1991. DOI: 10.1016/0747-5632(91)90002-i 9

[136] Ming-Zher Poh, Tobias Loddenkemper, Nicholas C. Swenson, Shubhi Goyal, Joseph R. Madsen, and Rosalind W. Picard. Continuous monitoring of electrodermal activity during epileptic seizures using a wearable sensor. In *Engineering in Medicine and Biology Society (EMBC), Annual International Conference of the IEEE*, pp. 4415–4418, 2010. DOI: 10.1109/iembs.2010.5625988 7

[137] Maja Pusara and Carla E. Brodley. User re-authentication via mouse movements. In *Proc. of the ACM Workshop on Visualization and Data Mining for Computer Security*, pp. 1–8, 2004. DOI: 10.1145/1029208.1029210 48, 54

[138] Robert W. Reeder and Roy A. Maxion. User interface defect detection by hesitation analysis. In *Dependable Systems and Networks, DSN. International Conference on*, pp. 61–72, IEEE, 2006. DOI: 10.1109/dsn.2006.71 48, 54

[139] N. H. Riche. Beyond system logging: Human logging for evaluating information visualization. *BELIV*, 2010. 3

[140] B. W. Roberts and W. F. DelVecchio. The rank-order consistency of personality traits from childhood to old age: A quantitative review of longitudinal studies. *Psychological Bulletin*, 126:3–25, 2000. DOI: 10.1037/0033-2909.126.1.3 14

[141] Julian B. Rotter. Generalized expectancies for internal vs. external control of reinforcement. *Psychological Monographs: General and Applied*, 80(1):1, 1966. DOI: 10.1037/h0092976 13, 29, 69

[142] Gillian Rowe, Jacob B. Hirsh, and Adam K. Anderson. Positive affect increases the breadth of attentional selection. *Proc. of the National Academy of Sciences*, 104(1):383–388, 2007. DOI: 10.1073/pnas.0605198104 4

[143] Batul Saati, May Salem, and Willem-Paul Brinkman. Towards customized user interface skins: Investigating user personality and skin colour. *Proc. of HCI*, 2:89–93, 2005. 14

[144] Robert E. Schapire. The strength of weak learnability. *Machine Learning*, 5(2):197–227, 1990. DOI: 10.1109/sfcs.1989.63451 62

[145] Andrew Sears and Ben Shneiderman. Split menus: Effectively using selection frequency to organize menus. *ACM Transactions on Computer-Human Interaction (TOCHI)*, 1(1):27–51, 1994. DOI: 10.1145/174630.174632 78

[146] Edward Segel and Jeffrey Heer. Narrative visualization: Telling stories with data. *Visualization and Computer Graphics, IEEE Transactions on*, 16(6):1139–1148, 2010. DOI: 10.1109/tvcg.2010.179 27

[147] Alexander J. Shackman, Issidoros Sarinopoulos, Jeffrey S. Maxwell, Diego A. Pizzagalli, Aureliu Lavric, and Richard J. Davidson. Anxiety selectively disrupts visuospatial working memory. *Emotion*, 6(1):40, 2006. DOI: 10.1037/1528-3542.6.1.40 4

[148] Ben Shneiderman. The eyes have it: A task by data type taxonomy for information visualizations. In *Visual Languages. Proceedings, IEEE Symposium on*, pp. 336–343, 1996. DOI: 10.1016/b978-155860915-0/50046-9 40, 77

[149] Pete B. Shull, Wisit Jirattigalachote, Michael A. Hunt, Mark R. Cutkosky, and Scott L. Delp. Quantified self and human movement: A review on the clinical impact of wearable sensing and feedback for gait analysis and intervention. *Gait and Posture*, 40(1):11–19, 2014. DOI: 10.1016/j.gaitpost.2014.03.189 xi

[150] Erin Solovey, Paul Schermerhorn, Matthias Scheutz, Angelo Sassaroli, Sergio Fantini, and Robert Jacob. Brainput: Enhancing interactive systems with streaming fNIRS brain input. In *Proc. of the SIGCHI Conference on Human Factors in Computing Systems*, pp. 2193–2202, ACM, 2012. DOI: 10.1145/2207676.2208372 67, 78

[151] Ben Steichen, Giuseppe Carenini, and Cristina Conati. User-adaptive information visualization: Using eye gaze data to infer visualization tasks and user cognitive abilities. In *Proc. of the International Conference on Intelligent User Interfaces*, pp. 317–328, ACM, 2013. DOI: 10.1145/2449396.2449439 10, 44, 62, 76

[152] Ben Steichen, Cristina Conati, and Giuseppe Carenini. Inferring visualization task properties, user performance, and user cognitive abilities from eye gaze data. *ACM Transactions on Interactive Intelligent Systems (TIIS)*, 4(2):11, 2014. DOI: 10.1145/2633043 10

[153] Melanie Swan. Emerging patient-driven health care models: An examination of health social networks, consumer personalized medicine and quantified self-tracking. *International Journal of Environmental Research and Public Health*, 6(2):492–525, 2009. DOI: 10.3390/ijerph6020492 xi

[154] Melanie Swan. Sensor mania! the internet of things, wearable computing, objective metrics, and the quantified self 2.0. *Journal of Sensor and Actuator Networks*, 1(3):217–253, 2012. DOI: 10.3390/jsan1030217

[155] Melanie Swan. The quantified self: Fundamental disruption in big data science and biological discovery. *Big Data*, 1(2):85–99, 2013. DOI: 10.1089/big.2012.0002 xi

[156] James J. Thomas and Kristin A. Cook. Illuminating the path: The research and development agenda for visual analytics. *IEEE Computer Society Press*, 2005. 39, 47, 67, 77, 78

[157] Dereck Toker, Cristina Conati, Giuseppe Carenini, and Mona Haraty. Towards adaptive information visualization: On the influence of user characteristics. In *User Modeling, Adaptation, and Personalization*, pp. 274–285, Springer, 2012. DOI: 10.1007/978-3-642-31454-4_23 xii, 10, 12

[158] Dereck Toker, Cristina Conati, Ben Steichen, and Giuseppe Carenini. Individual user characteristics and information visualization: Connecting the dots through eye tracking. In *Proc. of the SIGCHI Conference on Human Factors in Computing Systems*, pp. 295–304, ACM, 2013. DOI: 10.1145/2470654.2470696 xii, 10, 12, 13

[159] Archie Tse. *The New York Times*, 2016. 75

[160] Edward R. Tufte. *The Visual Display of Quantitative Information*. Graphics Press, 1983. DOI: 10.2307/2150427 xii

[161] Barbara Tversky. Functional significance of visuospatial representations. *Handbook of Higher-Level Visuospatial Thinking*, pp. 1–34, 2005. DOI: 10.1017/cbo9780511610448.002 10

[162] Barbara Tversky, Maneesh Agrawala, Julie Heiser, P. U. Lee, Pat Hanrahan, Doantam Phan, Chris Stolte, and M. P. Daniele. Cognitive design principles: From cognitive models to computer models. In L. Magnani, Ed., *Model-Based Reasoning in Science and Engineering*, pp. 1–20, King's College, 2006. 10

[163] Susan VanderPlas and Heike Hofmann. Spatial reasoning and data displays. *IEEE Transactions on Visualization and Computer Graphics*, 22(1):459–468, 2015. DOI: 10.1109/tvcg.2015.2469125 10

[164] Vladimir N. Vapnik. *Statistical Learning Theory*. Wiley-Interscience, 1998. 53

[165] Maria C. Velez, Deborah Silver, and Marilyn Tremaine. Understanding visualization through spatial ability differences. In *IEEE Visualization*, pp. 511–518, IEEE, 2005. DOI: 10.1109/vis.2005.108 xii, 4, 10, 12, 15, 20

[166] Kim J. Vicente, Brian C. Hayes, and Robert C. Williges. Assaying and isolating individual differences in searching a hierarchical file system. *Human Factors: The Journal of the Human Factors and Ergonomics Society*, 29(3):349–359, 1987. DOI: 10.1177/001872088702900308 xii, 10, 15, 20

[167] Fernanda B. Viegas, Martin Wattenberg, Frank Van Ham, Jesse Kriss, and Matt McKeon. Manyeyes: A site for visualization at internet scale. *IEEE Transactions on Visualization and Computer Graphics*, 13(6), 2007. DOI: 10.1109/tvcg.2007.70577 xi

[168] Emily Wall, Leslie M. Blaha, Lyndsey Franklin, and Alex Endert. Warning, bias may occur: A proposed approach to detecting cognitive bias in interactive visual analytics. In *IEEE Conference on Visual Analytics Science and Technology (VAST)*, pp. 104–115, 2017. DOI: 10.1109/vast.2017.8585669 4, 76

[169] Emily Wall, Leslie M. Blaha, Celeste Lyn Paul, Kristin Cook, and Alex Endert. Four perspectives on human bias in visual analytics. In *Cognitive Biases in Visualizations*, pp. 29–42, Springer, 2018. DOI: 10.1007/978-3-319-95831-6_3 4

[170] C. Ware. *Information Visualization: Perception for Design*, vol. 22. Morgan Kaufmann, 2004. xii

[171] Colin Ware. *Visual Thinking: For Design*. Morgan Kaufmann, 2010. xii

[172] H. Gilbert Welch. Overdiagnosis and mammography screening. *BMJ*, 339, 2009. DOI: 10.1136/bmj.b1425 20

[173] H. Gilbert Welch and William C. Black. Overdiagnosis in cancer. *Journal of the National Cancer Institute*, 102(9):605–613, 2010. DOI: 10.1093/jnci/djq099 20

[174] Leanne S. Woolhouse and Rowan Bayne. Personality and the use of intuition: Individual differences in strategy and performance on an implicit learning task. *European Journal of Personality*, 14(2):157–169, 2000. DOI: 10.1002/(sici)1099-0984(200003/04)14:2<157::aid-per366>3.0.co;2-1 9

[175] Yei-yu Yeh and Christopher D. Wickens. Dissociation of performance and subjective measures of workload. *Human Factors*, 30(1):111–120, 1988. DOI: 10.1177/001872088803000110 4

[176] Ji Soo Yi. Implications of individual differences on evaluating information visualization techniques. In *Proc. of the BELIV Workshop*, 2010. xii, 3, 10, 15

[177] Ji Soo Yi, Youn ah Kang, John T. Stasko, and Julie A. Jacko. Toward a deeper understanding of the role of interaction in information visualization. *IEEE Transactions on Visualization and Computer Graphics (TVCG)*, 13(6):1224–1231, 2007. DOI: 10.1109/tvcg.2007.70515 77

[178] Yingbing Yu. Anomaly detection of masqueraders based upon typing biometrics and probabilistic neural network. *Journal of Computing Sciences in Colleges*, 25(5):147–153, 2010. 48, 54

[179] Qiyu Zhi, Alvitta Ottley, and Ronald Metoyer. Linking and layout: Exploring the integration of text and visualization in storytelling. In *Computer Graphics Forum*, 38:675–685, Wiley, Online Library, 2019. DOI: 10.1111/cgf.13719 27

[180] Michelle X. Zhou and Steven K. Feiner. Visual task characterization for automated visual discourse synthesis. In *Proc. of the SIGCHI Conference on Human Factors in Computing Systems (CHI)*, pp. 392–399, ACM Press/Addison-Wesley Publishing Co., 1998. DOI: 10.1145/274644.274698 77

[181] Caroline Ziemkiewicz, R. Jordan Crouser, Ashley Rye Yauilla, Sara L. Su, William Ribarsky, and Remco Chang. How locus of control influences compatibility with visualization style. In *Visual Analytics Science and Technology (VAST), IEEE Conference on*, pp. 81–90, 2011. DOI: 10.1109/vast.2011.6102445 xii, 10, 13, 29

[182] Caroline Ziemkiewicz and Robert Kosara. Preconceptions and individual differences in understanding visual metaphors. In *Computer Graphics Forum*, 28:911–918, Wiley, Online Library, 2009. DOI: 10.1111/j.1467-8659.2009.01442.x xii, 10, 12, 14, 20, 25, 72

[183] Caroline Ziemkiewicz, Alvitta Ottley, R. Jordan Crouser, Krysta Chauncey, Sara L. Su, and Remco Chang. Understanding visualization by understanding individual users. *Computer Graphics and Applications, IEEE*, 32(6):88–94, 2012. DOI: 10.1109/mcg.2012.120 15

[184] Caroline Ziemkiewicz, Alvitta Ottley, R. Jordan Crouser, Ashley Rye Yauilla, Sara L. Su, William Ribarsky, and Remco Chang. How visualization layout relates to locus of control and other personality factors. *IEEE Transactions on Visualization and Computer Graphics*, 19(7):1109–1121, 2012. DOI: 10.1109/tvcg.2012.180 xvii, 13

[185] Caroline Ziemkiewicz, Alvitta Ottley, R. Jordan Crouser, Ashley Rye Yauilla, Sara L. Su, William Ribarsky, and Remco Chang. How visualization layout relates to locus of control

and other personality factors. *Visualization and Computer Graphics, IEEE Transactions on*, 19(7):1109–1121, 2013. DOI: 10.1109/tvcg.2012.180 xii, 4, 6, 40, 43, 52, 58, 73

[186] Coraline Ziemkiewicz and Robert Kosara. The shaping of information by visual metaphors. *Visualization and Computer Graphics, IEEE Transactions on*, 14(6):1269–1276, 2008. DOI: 10.1109/tvcg.2008.171 73

Author's Biography

ALVITTA OTTLEY

Alvitta Ottley is an Assistant Professor in the Department of Computer Science & Engineering at Washington University in St. Louis. She also holds a courtesy appointment in the Department of Psychological and Brain Sciences. Professor Ottley's research lies in the areas of data visualization and human-computer interaction. She uses interdisciplinary approaches to solve problems such as how best to display information for effective decision-making and how to design human-in-the-loop visual analytics interfaces that are more attuned to the way people think. Her prior research has focused on designing visualizations to support decision-making for non-experts. Her work has also made significant advancements in identifying individual characteristics (e.g., personality traits and cognitive abilities) that modulate behaviors with visualization designs and using machine learning techniques to model human interactions with visualization tools. She and her group continue to leverage this expertise to promote a symbiotic relationship between humans and machines. Professor Ottley earned her Master's and doctorate degrees at Tufts University under the supervision of Dr. Remco Chang. She was the recipient of an NSF CRII Award in 2018 and currently serves on the program committees for the leading conferences in visualization and human-computer interaction: IEEE VIS and ACM CHI.

Printed in the United States
by Baker & Taylor Publisher Services